INGENIOUS

INGENIOUS

The Unintended Consequences of Human Innovation

PETER GLUCKMAN · MARK HANSON

Harvard University Press

Cambridge, Massachusetts
London, England

2019

Printed in the United States of America

First printing

Library of Congress Cataloging-in-Publication Data
Names: Gluckman, Peter D., author. | Hanson, Mark A., author.
Title: Ingenious : the unintended consequences of human innovation / Peter Gluckman and Mark Hanson.
Description: Cambridge, Massachusetts : Harvard University Press, 2019. | Includes bibliographical references and index.
Identifiers: LCCN 2019014147 | ISBN 9780674976887 (cloth)
Subjects: LCSH: Human evolution. | Technological innovations—Health aspects. | Civilization, Modern—Health aspects.
Classification: LCC GN281.4 .G58 2019 | DDC 599.93/8—dc23
LC record available at https://lccn.loc.gov/2019014147

CONTENTS

INGENIOUS

INTRODUCTION

We were warned that this would not be a trip to the zoo. We were short of breath—not being acclimatized yet to the thin air at 9,500 feet above sea level. We had struggled through a steamy rain forest for three hours to reach this spot, sometimes sliding about on the path and often scrambling over tree roots. Our trousers were tucked into our socks to foil red ants, and our hands were sweating in the gloves we needed to grasp onto the spiky vegetation.

At the front of our group, a guide cut his way through vines and bamboo stems with a machete. Just when we thought we could go no deeper into the impenetrable jungle, we met three trackers. We put down our packs, took out our cameras, and followed them quietly. Suddenly there was a crashing sound above, and we looked up to see a female gorilla staring down at us.

Here, at the triple border between Rwanda, Congo, and Uganda, is the last refuge of the mountain gorilla. We moved forward cautiously and stood at the edge of a small depression. Imperiously, in the middle of it, sat a thirty-four-year-old silverback, Agashya, head of his harem. One of Agashya's sons, perhaps eighteen years old, nursed a large shoulder wound. He had fought his father for dominance in the troop, and he had clearly lost. Agashya kept a close eye on the other two younger adult males, also his sons, in the troop, and watched carefully over his harem of six adult females. They were in the bamboo, lying on the ground, grooming, or being pestered by their infants who were doing somersaults or engaging in mock battles. We knew that we must try to stay at least a few yards away from our primate cousins, but they made that impossible. Females returned from the trees, adolescents sauntered among us, and one of the younger silverbacks stood on his hind legs to grab a branch nearby.

The gorillas did not seem frightened or surprised to see us. It was us who became tense, holding our breath, uncertain of our place. This was their home and we were the intruders, high in the misty mountains of Volcanoes National Park.

We had been told not to look the gorillas in the eye because they might find it threatening, but their faces seemed so full of expression that it was impossible not to make eye contact. We were allowed to stay for an hour, mesmerized by the experience of being so close to a species so similar to us, and yet so different.

The ranger quietly told us Agashya's story. The harem had previously been that of another silverback, who had died. Rather than dispersing, the females stayed together. After some time, Agashya turned up, having crossed the border from Congo. The females had become habituated to humans over many years, and, remarkably, Agashya appeared to learn from his new wives that humans, at least those who came with rangers and guides, were not threatening.

About 98 percent of human DNA is identical to that of the gorilla. A small genetic difference has profound effects on what makes a gorilla a gorilla and a human a human. But how different are we really? Visiting their mountain habitat for that hour emphasized our similarities more than our differences. The gorillas obviously had a well-defined social structure and engaged in some complex form of communication, just as we exchanged meaningful looks with each other as we sat observing them. And there were clearly very different personalities: on the side of the hill, arms crossed in a defiant pose, one large female sat away from the group, looking antisocial. She is often in that mood, said our guide. But were we correct in attributing human emotions and feelings to these gorillas? Were they wondering about us in the same way that we were wondering about them?

Thinking about how similar gorillas are to us, and yet how enormously different, raises the age-old question: What makes us what we are? This question has occupied human minds probably for as long as we have been self-aware. To answer it, we must examine our nature—as many have done, from the earliest shamans to the discourses of every subsequent culture's theologians and philosophers.

The question of what makes us what we are is not only a metaphysical exercise, though. It can be addressed scientifically. This was one of the questions Charles Darwin explored in an argument extending across his three great books: *The Origin of Species, The Descent of Man,* and *The Expression of the Emotions in Man and Animals.* A well-known diagram in one of his early notebooks, enigmatically headed "I think," shows his idea of a single tree of life, by which all living and extinct organisms, from bacteria to dinosaurs, plants, insects, fishes, birds, mammals, and even humans, share a common ancestor.[1] Darwin saw no need to invoke a distinct creation of humans.[2] The mechanisms of inheritance of variations through natural selection, he suggested, were all that was necessary to explain the great diversity of the branches of the tree of life. Humans are only one small, albeit highly successful, branch.

Important to Darwin's emergent thinking were his visits to Regent's Park Zoo in London. He was fascinated by the emotional expressions of Jenny, the orangutan.[3] In 1838 he climbed into a cage with her to study her reactions and emotions. Indeed, the relationship of the gorilla to humans was the subject of intense battles over Darwin's great idea—namely, that there was a continuity of life and that all animals, including humans, were descended from a common ancestor.[4]

Through the course of evolution, human nature—and here we mean something much more than just our genetic makeup—and gorilla nature became wonderfully adapted to our respective environments. We share common ancestors that lived in sub-Saharan Africa some nine million to thirteen million years ago; we are cousins. But there the similarities end. Their ancestors followed an evolutionary path in West Africa, leading to the apes. Ours followed a path in what is now the Rift Valley to the east, leading to the origin of hominins.

Mountain gorillas' environment has been squeezed and is now limited to a vulnerable niche in the central massif of Africa. The family group we met are secure in their habitat because humans have allowed them to be so. We paid thousands of dollars to visit them and then retreated to the safety of our camp, and after that to our homes thousands of miles away. The gorilla family would not survive even a few miles outside their traditional habitat; they are only a few miles from the Democratic Republic of Congo, where poaching is rife. They have little recourse to remedy if one of them becomes ill or injured. If the supply of their food plants diminishes, they will face starvation. Gorillas have been seen to use some simple tools—such as a stick to judge the depth of water in a pool or to open a fruit—but they hardly have mobile phones, cars, and automatic weapons. They could never threaten our survival to any great extent, certainly not in the way we threaten theirs. We have created a niche for ourselves, too, but ours stretches across almost the entire planet.

Stories of our origins are well known, yet one of the most fundamental components of what makes us what we are has not been adequately appreciated, certainly not in terms of its contemporary implications. The ability to continually innovate and progressively change our environment is unique to humans. One only need look at the difference between our technology and that of gorillas. Gorillas' lifestyle has changed relatively little, if at all, over the last

100,000 years. But ours has changed stupendously—notably through our creation of increasingly complicated technologies. Is there something about our fundamental ingenuity that set humans, way back in evolutionary time, quite apart from all previous and future other species? Is our technological achievement not only quantitatively but qualitatively different?

Engaging with questions like these, we might worry that we are in danger of slipping back into a pre-Darwinian set of beliefs that saw humans as exceptional, not subject to the biological origins and constraints of other species—created separately, as maintained by Darwin's theistic critics. Does this book aim to challenge established evolutionary thinking? No, emphatically it does not: the fields of evolutionary biology and evolutionary psychology, which grew out of Darwinian theory, have produced thousands of pages of scholarly work on how human technological skills evolved. It is not our aim to challenge that work. Instead, we aim to build on that to understand the implications of our ingenious capacity for developing technologies. Our hypothesis is that this ingenuity is so central to human nature, and so fundamental to humanity's success in Darwinian terms, that our technological achievements effectively define and become our nature. In other species, biological and evolutionary responses to changes in the environment are usually slow, as new genetic constitutions more capable of survival radiate through populations. We do not wait that long. Moreover, while other species make changes to their environment in order to survive, we do so continually for many other reasons than just survival, most notably over recent generations for purposes of comfort, leisure, and to support an increasingly urban, energy expensive and consumerist, market-driven lifestyle.

Over the last few thousand years, at an accelerating rate, humans have developed and used technology to respond to their changing environments, which has led to even further changes to the environment, to our apparent advantage. Our ability to develop technologies, learn and communicate about them, and then redevelop them, is fundamental to human nature. So fundamental, in fact, that this ingenuity is, effectively, human nature. This is an idea with important implications.

Evolution is a game of survival. The gorillas in Rwanda, in their circumscribed environmental niche, are dangerously close to losing. In contrast, this game has been spectacularly won by humans. We have won so consistently, and for so long, because we have not let the challenges posed by nature—environmental change and competition with other species—gain the upper hand.

We will explore how humans have been able to change nature, and what some of the consequences have been. We know that our ingenuity brings benefits: from the simplest things like shelter to keep us safe and warm and a supply of food, to the most advanced forms of communication or medicine. But this innovation also has negative consequences: climate change, environmental degradation, and loss of biodiversity. The more we examine the consequences, the greater the range of unintended challenges we find arising from our technological ingenuity. To provide nourishment for a growing population, for example, humans developed methods to process and preserve food; but our easy access to today's energy-dense foods contributes to obesity. To protect ourselves from dangerous pathogens, we embraced cleanliness and invented antibiotics, which have led to rising rates of autoimmune diseases and antibiotic-resistant bacteria. More recently, our growing dependence on the internet and social media has been linked to troubled mental health and declining social cohesion. The applications of artificial intelligence, or AI, might undermine the very intelligent ingenuity which has enabled us to live as we do. We are not only changing our world, we are changing ourselves.

In this book we will consider the question of whether we can continue to meet the challenges our ingenuity has unintentionally created. We will start first by considering how humans evolved to be what we are now, before going on to discuss the unanticipated consequences of our distinctive evolution. What makes us what we are is the interplay between our evolved and inherited biology on one hand and our abilities to communicate complex ideas, to learn, and to make things on the other. We first explore these dimensions before, in the middle of the book, exploring how our ingenuity is changing us and then discussing the implications of these changes.

In Chapter 2 we examine the evolutionary imperatives of survival and reproduction. In Chapter 3 we consider how various forms of biological inheritance contribute to our story. Chapter 4 concerns a different form of evolution which sets us apart from other species—our cultural evolution. Then we use this information to consider how our ingenuity has changed our lives—from hunter-gatherers, to agriculturalists and settled living in Chapter 5 and to the impact of life in cities in Chapter 6. Then we turn in Chapter 7 to a new world where we live increasingly—online. It is this new technological world that represents the pinnacle of our cultural evolution to date, but in this world there are increasingly loud warnings about its harmful effects. In Chapter 8 we consider the unanticipated and increasingly threatening costs of our ingenuity. In our view we have important and urgent decisions to make and so in Chapter 9 we look to the future to identify a pragmatic path ahead—one that might allow our species and our societies to thrive in an increasingly complex and technological world.

But we will start, in Chapter 1, in a place where human survival is challenged even today—the Australian outback.

1 | THE OUTBACK

It was midday, the sun almost directly overhead. The car's thermometer registered the outside temperature as 34°C, only a small fraction lower than human blood temperature. There was no real shade. Thankfully, we had air-conditioning inside the car and an electric fridge in the back. We also had an awning attached to the side of the car and when we stopped, we pulled it out and sat under it, relieved to be taking a break from the bumpy drive in Australia's Northern Territory.

The landscape was remarkably uniform, somewhat dull. Occasionally we saw a group of wallabies loping along, or an eagle drop off from a bare branch and flap languidly away. Bushes were spaced evenly as far as we could see in every direction. They looked as if they were planted by an obsessed gardener trying to create an orderly display in an uninhabited rugged landscape. But the bushes themselves divided up the land. Each drew just enough moisture from the soil to survive in the dry season. Only when older bushes perished, or were stripped of their leaves, was there room for younger ones to thrive. Competition was fierce, even between members of the same species, in this place where resources were limited.

The bushes were not the only well-ordered features of the landscape. In every direction we could see termite mounds, substantial structures up to three meters high, made of compacted red earth glued together with the saliva of millions of insects. Like the bushes, the mounds are equally spaced, as if the colonies of termites negotiated territories. But they have another feature, unique to the mounds in this part of Australia. Whereas termite mounds in more temperate parts of the continent, and in Africa and South America, are roughly conical in shape, here they have two flat sides, rather like huge sand castles that have been pressed between the hands of a gigantic child.[1] Even more extraordinary is that the massive, almost two-dimensional mud castles are all oriented in the same way, mile after mile through the Australian outback.

Astonishingly, the mounds are accurately oriented north-to-south. It would be possible to navigate through the country just by following the direction indicated by the termite mounds, something that must have given some comfort to early explorers of the region. Every fifty yards or so across this landscape stands a direction marker, pointing north to the mouth of the Adelaide River and the Timor Sea. It's no wonder they are known as *magnetic* termite mounds.

We only needed to spend a few days here to learn important lessons about survival—for example, how the temperature varies over the day. At that time of year, the cloudless sky means that the temperature during the night drops dramatically. After a night shivering in our sleeping bags, we were glad to see the sun rising over the horizon and to stand in its warming rays while making breakfast. By mid-morning, temperatures rose and we shed our sweaters and jackets. As the sun sank back down in the late afternoon, we began to feel the chill and layered our clothing back on again.

Compass Mounds

How do other animals cope with these extremes in temperature without the ability to change clothing? One answer lies in the termite mounds. How is their construction determined, given that termites don't have compasses? The changing shadows cast by the mounds over the course of the day offer the answer. One flat side of each mound faces east, which maximizes the warming effect of the rising sun. At noon, when (in the Southern Hemisphere) the sun beats down from the north, its heating effect on the interior of the mound is minimized by the narrow northerly edge the mound presents to the sun. Then, in the evening, the rate of cooling inside the mound is minimized by the sun warming the flat west face of the structure. The mounds, that is, are constructed to minimize variation in temperature.

It's easy to imagine how, over many generations, the continually replicating colonies of termites achieved by trial and error a structure that keeps temperature variations inside to a minimum. Too much east–west construction, and the interior gets too hot in the middle of the day; not enough north–south extension of the flat sides, and it takes too long to warm up in the morning and cools down too quickly in the evening. In one still-growing mound—many of these mounds are hundreds of years old—a little excessive earth added in the east–west direction can be detected. The termites countered it with a little more extension in the north–south direction. Optimal conditions for the termite society within are maintained by orienting the mound north–south.

The termites have constructed a niche that allows them to flourish in their environment. The mounds' interiors have a complex architecture of chambers and passages that accommodate as many as several million inhabitants per mound. Each mound's intricate society has a queen, workers, and soldiers. Some termites leave the mound to forage for grass and other dry vegetation, which they bring back to be kept in special storage cells. This will be an essential food supply when the plain floods, as is common in the wet season between November and April. Safe in their castle above the water level, the termites can survive until the plain dries out again. Some termites never leave the mound and have adaptations for this life inside. Their skins and external skeletons are thinner—so thin, in fact, that they are almost transparent. Hidden from predators inside the mound, they do not need camouflage to resemble the grassland.

Over thousands of generations, the termites' bodies and behavior evolved, shaped by the force that Charles Darwin termed *natural selection*. The behaviors that led to construction of the best mounds favored the survival and reproduction of certain termites. Each new generation of termites may have had to learn some behaviors, but they could not have invented the entire strategy anew. If we were to transport a colony of ants from a European garden to this part of the Northern Territory, the European ants would not survive, not having evolved any mound-building behavior. They would be too hot in the day and too cold at night. In the survival game, they would rapidly lose when faced with this change in environment. The concept of natural selection led Darwin to his idea of the evolution of species. But it took a voyage to the other side of the world to help him formulate it.

The Doctor's Son and His Pigeons

As a child, Charles Darwin was fascinated by nature. It was not obvious how this would lead to a career, so he was encouraged to enroll in medical school in Edinburgh and follow in his physician father's footsteps. He soon abandoned the course, much to his father's displeasure, but while in Edinburgh he became deeply engaged with the scientific culture of the day. He was especially influenced by Robert Grant, an early transmutationist and a disciple of Lamarck and his study of small marine organisms.[2] Another career option, in line with Victorian ideas about acceptable lines of work for young men from respectable families, was to study theology. Supposedly engaged in this in Cambridge, Darwin instead spent his time collecting beetles and studying natural history with some of the scientific luminaries of the time. From the Reverend John Stevens Henslow he learned botany, and from the Reverend Adam Sedgwick, geology. He became fascinated by the work of Charles Lyell, who had revolutionized geological theory with his claim that many geological features are the results of gradual forces rather than catastrophic events.[3]

Darwin was also inspired by the writings of the explorer and scientist Alexander von Humboldt, who had traveled widely in South America and made highly original observations about the interactions among plants, animals, and their environments.[4] Darwin decided that he should undertake such an expedition himself. With Henslow and other friends, he planned to sail to the Canary Islands, where he was certain he would find the diversity of nature he longed to study. Unfortunately, Henslow withdrew from the plan and, in any case, it was not easy for Darwin to find either the ship or the financial resources needed to undertake the expedition. He was in despair. Then, in late August 1831, he heard from Henslow of a captain who needed a companion to sail with him on a long global expedition: Robert FitzRoy of HMS *Beagle*.[5] It took some convincing for Darwin's father to fund this venture, but perhaps he had already resigned himself to the idea that his son would not become a country parson any more than he would be a doctor. The *Beagle* embarked just after Christmas 1831 on a five-year expedition, and soon Darwin was no longer just a gentleman-companion but on his way to becoming the most distinguished naturalist and biologist of his time.

Darwin began that voyage believing in the natural theology—the concept that every feature of the world is a manifestation of God's direct handiwork—that was central to his theological studies in

Cambridge.[6] Over time that belief was replaced by new ideas based on his geological, geographical, and biological observations. A cautious scientist, he was aware of the revolutionary implications of his ideas and recorded them in his notes, only gradually sharing them with friends.[7] It was not until twenty years later that he actually published his ideas, spurred by correspondence from the collector-naturalist Alfred Russel Wallace, about whom we will say more in Chapter 2 and who was about to publish his own parallel recognition of the mechanisms by which species form.[8] In 1859, Darwin's "one long argument," as he put it—*On the Origin of Species by Means of Natural Selection*—appeared in print. It set off considerable controversy and debate, which have never completely gone away. There has been debate, as well, over the relative contributions of Darwin and Wallace to the fundamental concept. Wallace himself, however, gave precedence to Darwin by recognizing that his ideas had been developing over a much longer period.[9]

Darwin realized that variations in features of individuals within a species endow some of them with advantages over others in terms of their ability to survive and reproduce under prevailing environmental conditions. Even if the environment is constant, not every member of a species is perfectly adapted to it. Degrees of adaptation, and thus individuals' chances of reproductive success, vary. To the extent that an advantageous variation is heritable, this variant is more likely to be present in the next generation. Darwin termed this *natural selection,* recognizing the parallel with the conscious or artificial selection that farmers and breeders use as they choose animals or plants whose characteristics they want to see in subsequent generations. Darwin was fascinated by variation, and as he was working through his great idea, he spent much time with pigeon fanciers and livestock breeders.[10] At one stage he had sixteen different breeds of fancy pigeons at Down House, his home in Kent. The naturally occurring variation in any feature or *trait* within a species or a breed was fundamental to Darwin's concept of evolution.

Returning now to the Australian termites, we can consider how their present-day characteristics evolved. Unlike the interventions of Darwin's breeders, these processes were gradual and played out under natural conditions, so that the characteristics—the so-called *phenotypes*—of the insects shifted gradually. Successive generations were made up of more favorable variants than the previous ones, until an adaptive match between the termites and their environment was achieved and an equilibrium established—as long as the environment did not change again. There may have been some dramatic events in terms of climate or sudden changes in predator numbers along the way to accelerate the shift in the termite phenotypes; if so selection pressures would have been greater under such circumstances. We can also imagine a gradual migration of the insects, some better adapted to new territories than others, causing particular species of termites to end up in different places. But evolution is never finished. For every species, there continues to be a dynamic interaction between the range of its anatomical and physiological characteristics and its environment, which is also never totally stable. For the termite, constructing the mound, its niche, to minimize the potentially threatening aspects of environmental variations has been a critical adaptive strategy.

Today we think of the three tenets of Darwin's theory of evolution—phenotypic variation, natural selection, and organic (or, in modern terms, genetic) inheritance—as so fundamental to life and so uncontroversial that it seems hardly worth noting them. Yet we have to remember that the emphasis placed on these components, and even their necessary inclusion in his theory, was questioned from the very outset. Darwin does not once use the word *evolution* in *The Origin of Species,* referring only to new species' ability to *evolve* from the complexities of nature.[11] He uses the verb form as he ponders a tangled bank of vegetation on the last, and uncharacteristically poetic, page of the book.[12]

Darwin himself had no modern understanding of inheritance; the concept of the gene was yet to emerge, and it would be another hundred years before the structure of DNA was discovered, opening up the true study of genetic inheritance. Furthermore, Darwin was open to the ideas of earlier generations of transmutationists, including his grandfather Erasmus Darwin.[13] He particularly respected Lamarck, who had argued that environmental influences in one generation could lead to acquired characteristics being inherited by the next.[14]

In Austria in 1859, the year when the first edition of *The Origin of Species* was published, the monk Gregor Mendel was cross-breeding variants of peas and formulating the principles by which certain characteristics are passed from one generation to the next. Mendel's work was published in an obscure journal, however, and was yet to be discovered by other evolutionists and the broader scientific community. It was only when it started to receive recognition early in the twentieth century that its significance was recognized and the science of genetics was born.[15]

We now know that the basic unit of biological inheritance is the gene, which is a segment of DNA. Humans have about 22,000 genes spread over 46 chromosomes in 22 pairs of chromosomes (we have two copies of each) plus our two sex chromosomes (two copies of the *X* chromosome in females and one *X* and one *Y* chromosome in males). Other apes, including the gorilla, have 48 chromosomes; at some stage in our evolution from a common ancestor with the other apes, two chromosomes fused into one.[16] But most of the DNA on a chromosome does not actually comprise the genes that lead to the instructions to make proteins (the traditional definition of a gene);

rather this DNA provides regulatory control over whether a particular gene is turned on or not and under what circumstances. This is what leads to the complex regulation of gene expression. Subtle variations between individuals in the DNA sequence within and around a gene are widespread and can lead to changes in gene regulation, and sometimes to changes in the protein structure, with consequences for the biological effect of the gene and protein it codes for. The sum total of genetic information in an individual is called a *genotype.* The total expression of gene function, structure, and behavior gives rise, along with other factors we shall consider later, to the phenotype.

Darwin developed his ideas without the benefit of understanding inheritance, but once the gene was discovered, population geneticists came to recognize the importance of genetic inheritance to Darwinian theory. What has been termed the *Modern Synthesis,* or *neo-Darwinism,* emerged gradually in waves of great intellectual thought in the 1930s, 1940s, and 1950s. Modern evolutionary theory is based on this synthesis, although as we will see there are still significant debates over the role of other evolutionary processes. At the greatest extreme, there has emerged a very gene-centric school of biology that regards genes alone as able to explain almost all biology. The scientists and members of the media who refer wrongly to a "gene for religion"—or for violence, arthritis, diabetes, dementia—are claiming an extreme genetic determinism that ignores much of biological reality.

Explorers' Footsteps

As we climbed back into our air-conditioned car and drove away from the termite mounds, we couldn't help thinking about the lives of the first European explorers and settlers who eked out an existence in this harsh part of the world. Frederick Henry Litchfield, after whom this national park is named, was part of the expedition, led by Boyle Finniss, which traveled overland from South Australia in 1864 in hopes of establishing a settlement in northern Australia.[17] The expedition, like others before it, was unsuccessful, and the death toll from undernutrition, insect-borne disease, and infections was high. Its members traveled more than 2,500 kilometers through some of the harshest terrain on earth, carrying everything they needed on horses, mules, and camels. Most of the animals died. Unlike the termites, the explorers and their animals had neither the evolved biology nor the capacity to establish a niche that would have enabled them to survive.

Less than 150 years later, as we bumped along the same road these early explorers took, we were grateful that our survival chances were much better. We had packed enough food to give us a varied diet for the trip, and had plenty of clean water. Before going out into the sun, we slathered on sunscreen with a high UV protection factor. We had antibiotics and first aid kits and insect repellents. Of course, we knew that if our car broke down and we were stranded in the outback for a prolonged period, we would likely perish; and that if one of us had the misfortune to be bitten by a king brown snake, we would be in serious trouble.[18] But we had a radio beacon we could activate as a signal for help via satellite, and emergency assistance would reach us by air in only a few hours. It occurred to us that, in the car with all our equipment and resources, we were in a kind of mobile niche for human survival.

Looked at this way, it may seem that this place is so inhospitable that humans would have only recently gained the ability to settle here. Nothing could be farther from the truth. Some of the paintings in rock shelters and the stone artifacts found in this region are over 40,000 years old, made by ancestors of the first-nation Australians who inhabited this part of Australia, historically called Arnhem

Land. Their ancestors had migrated from the Rift Valley in Africa, northeast into the Eastern Mediterranean, then down through Southeast Asia, using land bridges that connected many Indonesian islands. The continent of Australia was connected to New Guinea at that time, but how these people crossed the last sea gap between Southeast Asia and the Australian continent—whether accidentally or intentionally—we do not know.[19]

Arnhem Land may have been one of the first places these ancient migrants settled. Until the arrival of Europeans in the nineteenth century, first-nation Australians' way of life had not changed significantly over thousands of years. The rocky outcrops provided shelters from the heat or rain, and the grassland bush had plentiful game. Early rock art shows kangaroos, wallabies, and ostrich. The rivers provided plentiful water, and there was a wide range of tubers, grubs, and fruits to eat. These first-nation peoples had a fairly healthy diet, and were able to move around to prevent exhausting the resources of one place.

From what we know of more recent first-nation tribes in the Northern Territory, as documented by explorers, most moved around as groups, traveling light and stopping in the places that were best for a particular time of year. Like we did in our technologically complex car, they took what they needed with them so they could sustain themselves wherever they went. But unlike us, they took only simple things, like their spear-throwing sticks and fiber baskets, and they negotiated their existence in terms of what they found around them. They evolved many different languages as they spread over Australia.[20] And they maintained rich cultural variations, including in their storytelling traditions and complex social practices concerning passage through life, such as who could marry whom and rites associated with the transitions to adulthood or death. They had found a niche to inhabit.

Finding a Niche

Environments can vary over the course of many years. They can also fluctuate between years and seasons and even over days. This means that however well suited a species is to an environment, it must be able to cope with changes in that environment. Most species' physiology allows them to cope with very transient changes. For example, our sense of thirst is triggered when we become dehydrated, and our kidneys allow us to excrete water when we are overhydrated. These two complex processes are controlled by several hormones released in response to changes in the composition and volume of the blood, maintaining the appropriate level of hydration for cellular functions and the cardiovascular system to operate effectively.

Longer-term and persistent environmental change, which is our focus here, requires a different set of strategies. The species may move to a more acceptable environment—as many species of fish in the Pacific Ocean are doing today by shifting to different latitudes to cope with changing ocean temperatures.[21] Some species cope by altering their behavior. Today, in response to climate change, many flowers are opening earlier in the season and birds are changing their nesting habits. Other species, like termites, may be able to cope via *niche construction*—building and maintaining an effectively constant microenvironment.[22] And it is inherent in the evolutionary paradigm that some variants in the population may be more likely than others to survive in a changing environment—for example, in the famous finches Darwin observed in the Galápagos Islands, beaks changed shape across generations according to what foods were available, as different beak shapes are more suited to different types of seed.[23] If this range of strategies does not provide the species with an adequate capacity to adapt, then the species is in trouble.

Humans are a major cause of rapid environmental change to which many other species cannot adapt. Deforestation, intensive agriculture, livestock production, hunting, fishing, and now pollution and greenhouse gas emissions are all playing their roles in the precipitous decline in biodiversity across the world. This is not a recent phenomenon, though. Pre-human Australia's wide range of megafauna, including giant kangaroos and wombats, became extinct soon after humans arrived, as also happened later to megafauna in the Americas, Madagascar, and New Zealand when humans arrived.[24] What is new today is the scale and speed of the environmental changes and the effects we are producing on other species.

The ancient rock art of the first-nation Australian people provides some evidence of their adaptability. About 10,000 years ago, when rising sea levels brought the coastline nearer to their rock shelters, sea creatures—both mythical and real—began to appear in their paintings. Closer contact with the sea must have brought about some changes in culture and lifestyle. In addition to new sources of food, there would be new sources of both danger and inspiration. They needed to develop new technologies and new skills, such as sophisticated fishing traps, that could be passed from generation to generation.

These previously well-adapted first-nation peoples met severe challenges when their environment changed for the worse in the nineteenth century. This was due, not to famine, flood, or volcanic eruption, but to the arrival of Europeans with guns, alcohol, tobacco, and new diseases. These newcomers were determined to use their technologies to change the environment for their own benefit—clearing the bush, planting crops, grazing cattle, and setting up new communication systems. New niches were constructed that largely suited the invaders. While it may not be appropriate to romanticize the lives of the ancestral first-nation people, which at times must have been very harsh, nonetheless it must be noted that the conditions on reservations today are far worse in many ways. Descendants of those peoples now lead lives that are sedentary rather than nomadic, have unhealthy diets, and experience high levels of drug and alcohol abuse, violence, smoking, obesity, diabetes, and cardiovascular disease. Their life expectancy on average is ten years shorter than that of other Australians.[25] They are no longer living in a healthy niche.

A niche need not be just physical. It can also be biological—related to the attributes of the species, to the broader biotic ecosystem of other plants and animals, and to the species' behavioral characteristics. For this reason, not all species that would be physically able to inhabit a particular niche actually do so. The large predators of the African savanna offer a good example. The territories of lions and cheetahs usually do not overlap, and it is only when a pride of lions dies out that cheetahs move to occupy that pride's former territory (and only if some adjacent pride of lions does not expand to that territory first). Nature abhors a vacuum, as Aristotle first observed.[26] But nature doesn't like overcrowding and unnecessary competition, either.

Protecting Ourselves

The extremely dark skin of native Australian people is an adaptation to sun exposure, which they evolved or possibly retained from their African ancestors.[27] It reduces their risk of developing skin cancer. The early European settlers' light-colored skin left them defenseless against Australia's intense sun, and even today this part of the world has one of the highest rates of skin cancer. Awareness of the importance of sunscreen and protective clothing became widespread in Australia only in the past couple of decades. The risk of skin cancer has been increased by recent technological developments—including refrigeration, which causes atmospheric accumulation of chlorofluorocarbons (CFCs), which in turn damages the ozone layer. As this layer becomes thinner, the penetration of ultraviolet light to the ground increases. The widening hole has also raised the risk of skin cancer in the Western Hemisphere. The Montreal Convention of 1999, which led to the agreement to abandon the use of CFCs as refrigerants, remains a high point in international environmental agreements.

If protection from the sun through darker skin pigmentation was a useful evolved trait for our ancestors in the tropics of East Africa, and for those who migrated to the tropics of Australia, we might ask why it did not persist in Europeans, who were, after all, also descendants of the same early humans. The answer lies in continuing processes of biological evolution. Under the influence of ultraviolet radiation in sunlight, the skin of the human body synthesizes most of the active vitamin D the body requires. Having adequate levels of active vitamin D is essential for healthy, strong bones, and for growing the skeleton before and after birth. Deficiency of vitamin D in childhood is associated with rickets, a serious condition in which the bones become weak—notably in the legs, to the point that they do not support the weight of the body adequately and become bowed. Another danger, particular to young women, is that rickets affects the shape of the growing pelvis and can therefore make delivery of a baby difficult. In the absence of modern obstetrics and cesarean section, such obstructed labor can cause death of both mother and baby.[28]

As early humans migrated northward into Europe, their exposure to sunlight decreased, reducing the skin's natural production of active vitamin D. This would have increased the chances of rickets and reproductive problems. Individuals with lighter skin would have been able to produce more vitamin D, because more sunlight would have penetrated their skin, and therefore would have been more likely to be healthy and to have healthy pregnancies and children. So, naturally occurring variations in the genetic control of skin pigmentation that led to selection for less-dark skin, and thus to more active vitamin D production, were favored by natural selection. It is likely that there were also some spontaneous genetic mutations that reduced skin pigmentation in the northerly-migrating people, accelerating the evolution of light-skinned people. Ethnic differences in skin pigmentation are still evident today—in general, people living nearest to the equator have the darkest skins, and those who live nearest to the Arctic have the lightest.[29]

At the time when light-skinned Europeans, ill-equipped for the intense sunlight, were traveling north to Arnhem Land from southern Australia in the middle of the nineteenth century, events in parts of Europe were putting light-skinned people there at greater risk. One of the reasons many people had left Britain for the Antipodes at that time was to escape the poverty and poor living conditions in the country's industrialized cities. The movement of workers from impoverished rural communities into cities during the Industrial Revolution, starting in the late 1700s, led to an urgent need for cheap, high-density housing. The resultant tenement buildings were separated by narrow alleys that let in very little sunlight, which reached street level only after filtering through the smoke from factories and domestic fires. Photographs and medical records from the mid-1800s show that many children in these cities suffered from rickets. They were exposed to little sunlight at home—or at school, if they were lucky enough to attend it—and after about the age of ten they started work in mills and factories, which kept them totally out of the sun. In this respect, their parents or grandparents, who had worked as laborers on the land, led healthier lives. The cities of the Industrial Revolution might have provided a good niche for human occupation—at least in economic terms for the mill and factory owners—but they also brought other problems, as we will see in Chapter 6.

Vitamin D deficiency became widespread during the Industrial Revolution and persists today in many parts of the world. Modern urban life typically means living in apartments, moving about on public transport or in cars, and working indoors in offices. This way of life prevents many of us from getting adequate exposure to the sun. Europe's northern latitudes are especially problematic. In the summer months there, twenty minutes of exposure of about one-fourth of the body's skin to sunshine will maintain adequate active vitamin D levels; but in the winter there, even lying outside all day, naked, would not achieve these levels. In parts of the United Kingdom, for example, over 20 percent of pregnant women are vitamin D deficient. This has consequences not only for their own bones and cardiovascular systems but also for their children's health. Doctors there are now reporting cases of rickets in children again.

Darwin's Clock

After returning from the voyage of the *Beagle* and marrying his cousin, Emma Wedgwood, Darwin spent a few years in London sorting his many specimens from the expedition and sharing them with experts. He then moved to Down House in Kent, where he was to spend most of the rest of his life, punctuated by frequent trips to English spas to deal with a persistent stomach problem.[30] He continually engaged in correspondence with naturalists, adventurers, collectors, and animal breeders across the globe, requesting specimens and observations from others. He played host to many of the distinguished scientists of his age. More than a home with an office, Down House was his laboratory. He spent many years demonstrating his biological credentials with detailed work on barnacles—work that was to play a major role in his thinking about sex. But his experiments, often assisted by his growing number of children, extended to plants, seeds, rabbit skeletons, earthworms, and pigeons.

Many of his ideas about the origins of variations in animal and plant characteristics came from observing the programs used by breeders to produce particular sets of traits over relatively small numbers of generations, employing a process of artificial selection. Darwin came to realize that the natural processes of selection in the wild also generated variants and, over longer periods of time, produced related but different species in distinct ecosystems. Darwin had collected finch specimens in the Galápagos Islands, but it was only later that the great ornithologist John Gould recognized and told Darwin that these birds were actually a mix of different species, related but distinct.[31] Their variations suggested a dispersal of groups of finches at some time in the distant past to the different islands (and, as later research found, to different niches on the same island). Anatomical differences had arisen over many generations as naturally selected adaptations to the varying conditions on the islands. The best-known finding was that varying beak shapes seemed to be well adapted to different kinds of foods available on the individual islands.

Characteristics such as beak shape "breed true," meaning that they are reliably passed from one generation to the next. Such variations are usually thought to be genetic, and indeed, the DNA sequence has been shown to vary between the different types of birds. Multiple genes are involved in beak structure, so selection occurs over many generations to optimize it. But as the ornithologists Peter and Rosemary Grant have shown, adaptation can also occur rapidly when conditions of severe drought or rain diminish food supply and create strong selection pressure.[32] The advantage of having a beak shape particularly suited to the enduring food supply has a dramatic impact on survival. To the extent that this advantage is genetically determined, if the drought or other pressuring conditions persist over multiple generations, the proportion of the finch population featuring the drought-adapted beak shape rises progressively, and the species evolves. If, on the other hand, the weather pattern reverses, and a rainy year follows, those finches whose genes give them a beak shape more suitable for wet conditions are more likely to thrive. Thus, with alternating weather over successive years, the distribution of beak sizes in the finch population fluctuates. The Grants' observations support Darwin's thesis that even subtle variations in heritable phenotypes can have major effects on survivability and thus on evolution.

However, as we will see in Chapters 3 and 4, changing environmental conditions can also produce *epigenetic* effects that can change the way genes work by affecting the regulatory components of DNA. Because induced epigenetic changes often take place during early development, they can affect the characteristics of multiple individuals within a population much faster than changes in the genes themselves, which take many generations to become widespread throughout a population. Moreover, because epigenetic changes can be passed from one generation to the next (at least for one or two generations; see Chapter 3), they do not have to be produced afresh in each generation. Recent research shows that there are epigenetic differences between the strains of Darwin's finches across the Galápagos Islands, and even between birds living in more urban or more rural areas of the same island.[33] It looks as if the environmental changes on the islands, caused partly by human habitation and tourism, are affecting the variants of the finches on a much faster scale than expected. Is this rapid change between generations of finches the result of genetic selection in each generation, or of epigenetic change, or perhaps both? The jury is still out, but the verdict will have important implications for evolutionary thinking.

Natural selection, moreover, is only one part of the evolutionary process. The selected trait must be heritable and underpinned by genetic differences and thus sustained as a genetic change within the population. And in general, unless there is very strong selection pressure, this takes a number of generations. Phenotypic measures, such as beak length, are relatively crude, but where molecular measurements are available, there might be more subtle evidence of selection and evolution. Evolution is then seen to be associated with a shift in the genetic structure of the population.

There is now a growing literature demonstrating rather rapid natural selection and evolution in lizards, fish, birds, and mammals. For example, wild male guppies translocated from a high-predator to a low-predator environment can evolve to have increased bright coloration in as few as three generations, or one year. This bright coloration is sexually attractive to female guppies, so it offers a reproductive advantage (or, in evolutionary terms, a fitness advantage—*fitness* being the reproductive potential of the organism). But having this coloration also brings attention from predators, so it comes at the cost of greater survival risk to the male.[34] Native green lizards in Florida have faced recent competition from invading brown anole lizards. As a result, the native lizards have evolved over perhaps twenty generations to have bigger toepads and stickier scales. Able to survive on higher and thinner branches, they have abandoned the lower, thicker branches to the brown lizards.[35] In under fifty years, the color balance of tawny owls in Finland has shifted from grey to brown. The grey used to help them hide from predators in the winter, but as white winters have become less common in recent decades, the genes that induce brown coloration have been favored in natural selection.[36] Climate change causes pink salmon in Alaska to spawn earlier, and this has been linked to changes in gene frequency.[37] Finally, rapid evolution has also recently been demonstrated in mammals. House mice are normally sensitive to the poison warfarin, but Algerian mice are resistant. Typically, when mating occurs between very different species that separated several million years ago from a common ancestor, the hybrid offspring are sterile, but viable hybrid offspring have been produced by Algerian and house mice. The fact that the gene that confers warfarin resistance in the Algerian mice can now be found in house mice is clear evidence for ongoing evolution as this gene spreads through the mouse population, in this case responding to a human innovation.[38]

How rapidly can biological evolution occur in humans? In this context, *rapid* means over a few generations, and the answer depends on many factors, particularly three: (1) The level of genetic control over the trait being selected: Most traits involve multiple genes, and thus selection is very slow. (2) The impact of the variation on the survival and reproductive success of the individual: If the effect is very subtle, the variation will take many generations to be selected, if it is selected at all. (3) The magnitude of the environmental change, and its impact on the reproductive success of individuals exposed to the new environment: This will determine the selection pressure and thus its speed.

These factors can be illustrated in terms of another environmental challenge: low levels of oxygen in the air we breathe in mountainous areas. Certain characteristics of the Aymara people, who have lived on South America's high Andean Plateau for about 10,000 years, make them better-adapted to that environment than residents whose ancestors arrived there from Europe only a few hundred years ago. The Aymara people's flattened faces protect them against the effects of the cold, high-altitude wind. Their barrel-shaped chests give them larger lung volume than Europeans of the same height; they have more oxygen-carrying red blood cells and a greater density of capillaries in many of their tissues, enabling their blood to carry more oxygen and their circulatory system to deliver it more effectively to their organs. In these respects, they resemble other high-altitude peoples, such as the Tibetans of the Himalayas. These characteristics appear to be adaptations selected by evolutionary processes, because they are embedded in the people's genetic makeup—indeed, the genetic basis of these physiological differences is now understood. The same characteristics are apparent even in Aymara who now live at lower altitudes, such as in Santa Cruz in Bolivia. But the metabolic changes that allow Aymara to live at high altitude are quite different from those of the Tibetans, who have a different set of genetic changes affecting the way their blood cells carry oxygen to similarly allow them to live at high altitude.[39] This demonstrates a phenomenon known as *convergent* evolution, by which organisms can achieve similarly effective outcomes by quite different routes. We can further contrast the Aymara and the Tibetans to Han Chinese, who have been living at high altitudes in Tibet and Western China for only a few hundred years. The Han do not show these adaptive characteristics, and their health can be challenged as a result.

Other examples of relatively rapid evolutionary change concern metabolism and responses to heat or cold, which differ substantially between ethnic groups. The Inuit of Greenland are more able than Europeans to metabolize some of the fatty acids in foods.[40] The amount of fatty food consumed by the Inuit would be sickening, if not downright toxic, for Caucasian people. The Inuit traditional diet is high in fat from the blubber of seals and other arctic aquatic mammals and in fish. The oils in the bodies of fish have a high content of omega-3 fatty acids because they feed primarily on plant sources or, if predatory, on other fish.[41] These oils help the fish survive in very cold oceans—even at low temperatures, their tissues do not solidify like butter does in the refrigerator. In turn, mammals consume the fish and the Inuit eat both the fish and the mammals. Ethnic groups from high latitudes also have higher proportions of brown adipose tissue in their bodies. This type of fat appears brown or beige because it has a high density of the energy engines of cells, the mitochondria. And these are regulated specifically to burn energy to generate heat.

Each of these changes to the structure and function of the body has evolved over a relatively short timescale, well within 10,000 years, in various parts of the world. They have enabled us as a species to occupy a diverse range of niches, from the equator to the Arctic Circle, conferring useful adaptations to these disparate environments. The characteristics have become assimilated into the genomes of our different ethnic groups such that they are heritable and reasonably stable: Inuit who migrate to the equatorial rain forest would be worse than uncomfortable—it would be extremely difficult for them to survive there, and this would not improve in a few generations.

A number of regions of the human genome show evidence of accelerated rates of change relative to our ape cousins. Although the procedures involved are complex, it is statistically possible to distinguish between genetic change due to natural selection and what might be just random mutations within a population, or accidental shifts in gene frequency due to what is called *genetic drift*.[42] A number of genes appear to have been under positive natural selection over the last few thousand years in humans.[43] Many of these changes relate to genes associated with brain size, reproductive traits, the environment, our relation to infectious agents, or nutrition. Some aspects of our biology can evolve quickly when our living conditions are modified. The domestication of fire enabled us not only to keep warm but to cook our food, which makes it easier to digest and to absorb its nutrients rapidly in our gut. As a consequence, we developed shorter guts than our cousins the chimpanzees. And there was no longer selection pressure for massive jaws and jaw muscles, so human faces are much flatter than chimpanzees' faces. Our teeth are also different. With the change to softer foods, our mastication came to involve more grinding than tearing. Some research suggests that when human infants began to be fed on particularly soft foods in recent centuries, the reduced need for grinding movements led to an increased risk of malocclusion.[44]

We are beginning to see that, far from being constant since the time of our Paleolithic ancestors, several aspects of human biology have continued to show biological evolution. Some of these changes seem to have come about relatively slowly in populations as a whole; others may have occurred faster, over perhaps a few generations. Both types of variation could be subject to the natural selection envisaged by Darwin. By surviving—or not—in the population, they provide the substrate for evolution. But the game of survival can be tricky to play. We explore this further in Chapter 2.

2 | SURVIVAL

As Charles Darwin realized, natural selection operates to favor the survival of some members of a species over others. Natural processes give rise to variations in characteristics among members of a species, and these will make some members better adapted to their environment than others. Those that are better adapted will be more likely to breed successfully and pass on their characteristics to the next generation. Evolution sets the rules of the survival game: this game is the focus of this chapter.

But the environment changes, whether from place to place, between seasons, or from year to year. So any species must have strategies to survive in the face of such changes, and this requires the species to be able to change some aspects of its structure and its function to meet the challenges of both short-term and long-term environmental changes. Like all other species, humans must also have this ability. Our physiology evolved to allow us to cope with very short-term environmental changes. If we could not sweat, we would have a higher propensity to heat stroke and could not cope with the daytime temperatures in many parts of the world, such as the Australian outback we encountered in Chapter 1. If we could not shiver and use a particular type of fat, namely brown fat, in our bodies to generate heat, we would be more susceptible to hypothermia on a freezing day. Every organism has physiological mechanisms that enable it to adapt to environmental change. Species with extensive adaptive capacities are "generalist" species, whereas those with very limited adaptive capacities must occupy very specific niches. To understand the human condition, we need to explore these concepts in more depth.

The successful generalist species can survive and reproduce in many different environments. To do so, it relies on its adaptive physiology and behaviors, with its offspring spreading out to find new niches to occupy. But for most species there are limits on where they can thrive. Some can thrive only in the very specific environments for which they have evolved. The lion could not survive in the Arctic; the polar bear could not survive in the Maasai Mara National Reserve in Kenya. Indeed, most species have adapted through evolutionary processes to occupy rather specific niches, defined by both biological considerations and relations to other species in the environment. The koala needs the blue gum tree as its food source, and the panda needs particular species of bamboo. Many flowers have coevolved with pollinating insects—for example, an orchid discovered in Madagascar has a very long spur, and Darwin speculated that it must have evolved this structure to aid pollination by a moth with a very long proboscis. In Darwin's time, no such moth was known, but sure enough, it was discovered not long after his death.[1]

For most species, survival is part of a dynamic equilibrium. If the environment changes, the species must either move or adapt—or, like the termite, build a defense. There is no possible long-term outcome other than survival or extinction.

At the end of Chapter 1 we considered the responses of human populations to the level of the essential element oxygen in the air. Another essential requirement for life in all environments is water. For this reason astrophysicists investigate the environmental chemistry of other planets and celestial bodies to ascertain whether they contain water. If an environment lacks water, we cannot conceive of life there—or at least, to use the old phrase from Star Trek, "life as we know it." Without water, we cannot imagine growth, reproduction, movement, or response to stimuli. But life can exist for some time even without water, in the driest conditions. In the face of drought, some bacteria and algae divide to generate spores that can remain viable even in dry soil for years, to form reproducing organisms again if water returns. This is clearly a highly effective rule for survival: In the face of adversity, such organisms go into standby mode and wait for the problem to go away. Of course, if the drought is permanent, this strategy fails. The strategy of staying put must be based on the likelihood that conditions will improve, and this strategy works for those species that survived.

In more complex organisms such as plants, the strategy is more refined. Drought eventually dries out a plant to the point that its metabolism shuts down. Waxy leaves and hardened impermeable bark have evolved to limit water loss and prolong survival. Plants living in deserts have deep roots to capture moisture. Even if there is no water at all in the soil, there may still be ways to survive. The Atacama Desert in northern Chile is the driest desert on earth, with no recorded rainfall. But at a certain altitude, mist forms at night as the tiny amounts of water in the atmosphere condense in the cold. This gives the candelabra cactus an opportunity to exploit a niche that almost no other species can occupy. Its shallow roots serve only to anchor it, but the spines on its branches—the candelabra-like arms that inspired its name—allow it to extract sufficient moisture from the misty evening air.

Plants cannot move out of their environments in the face of a drought, but they can disperse their seeds via the wind, water, or the droppings of animals and birds that have eaten them.[2] This may allow their offspring to reach a more favorable environment. In this respect they find a suitable niche, albeit sometimes remotely, by the random transfer of seeds. If these germinate in favorable conditions, the plants may achieve prominence over competing species and alter the ecological diversity of the new environment. Gardeners are aware of the constant battle against weeds in their flower and vegetable beds, but occasionally one weed emerges that seems to take over completely. This is the case for Japanese knotweed, which was brought into England to enhance the collection in Kew Gardens in London during Victorian times.[3] Now it is the scourge of many gardens, spreading rapidly via its rhizomes, and it is backbreakingly difficult to uproot.

Another strategy for survival is the evolution of fire resistance. Anyone who has witnessed a fire spreading through a pine forest or the eucalyptus trees of the Australian outback knows how fast fire can move, aided by any wind and accelerated by the dry leaf and needle litter from these trees and the highly flammable resins or oils they secrete. These vaporize in the heat, enabling the blaze to leap from tree to tree with alarming speed. Yet pines and eucalyptus are extremely good at sprouting again after such a fire. Because bush and forest fires occur naturally at intervals, in the absence of human interference, the evolution of this fire-promoting yet fire-resistant trait has enabled such tree species to construct a very effective niche for themselves.

Migration

Many animals have alternative available strategies. For instance, some can relocate if the environmental conditions do not suit them. They might not be able to move far enough to avoid a drought, but if such droughts occur on a regular annual basis, they may have evolved a strategy to move at a certain time of the year. The annual migration of millions of wildebeest, zebras, and other herbivores at the start of the dry season to the Maasai Mara, in East Africa, is one of the most striking examples of such behavior. These species have evolved a pattern of migration that is so stereotyped that predators such as crocodiles are able to predict when the wildebeest will make a river crossing—and they will lie in wait (along with a multitude of well-heeled tourists with their cameras) so as to have a feast.[4]

Humans as a species are experts at migration of a less regular nature, and we started these migrations many millennia ago. By 1.5 million years ago, our ancestral species—*Homo erectus*—had already spread from Africa across Asia. Not only *Homo sapiens* but also other species of hominin evolved from *Homo erectus:* the Neanderthals in Europe, the Denisovans in Central Asia, and the dwarf hominin in Flores in Indonesia.[5] At least 100,000 years ago, the first of our own species left Africa for the Middle East, to be followed perhaps 80,000 years ago by a second wave of migration. By 60,000 years ago they had walked through the Middle East across South Asia, into Southeast Asia then made their way to Australia, where they became the ancestors of the first-nation people of Australia.

As *Homo sapiens* occupied Africa and then the European, Asian, and Australian continents, they continued to spread across the range of environments they could occupy. Perhaps 14,000 years ago, a small group crossed a land bridge between Siberia and Alaska and over the next 4,000 years occupied the Americas. As our ancestors learned to make rafts and then boats, some radiated from East Asia into the Pacific until, 800 years ago, New Zealand became the last significant land mass to be settled by our species. As *Homo sapiens* thus moved into a great variety of environments, ranging from the Subarctic to the Polynesian Islands, we adopted behaviors, technologies, and social structures that enabled our survival. And as we have seen in our examples of altitude acclimation and varying shades of skin color, our biology often changed to assist our survival.

Although Neanderthals coexisted with *Homo sapiens* for a while in Europe, and some of their genes survive in humans as a result of interbreeding, Neanderthals went extinct about 28,000 years ago. It is not entirely clear why this happened, but it is generally thought that humans displaced them from environments in which they could thrive and squeezed them into marginal environments they did not have the capacity to adapt to.

As human societies evolved into larger groups, and cities and states formed, conflict and slavery fueled other forms of migration. Examples are legion: The Bible has many stories of forced migration; African slaves were common in ancient Rome; Vikings invaded Scotland and England; Genghis Khan and the Mongols invaded Western Asia from Mongolia; the invasion of the Americas by Europeans was soon accompanied by the horrific slave trade from West Africa; and today war and conflict continue to create reasons for migration, perhaps more than ever before.

When humans are subjected to political oppression, economic ruin, or physical aggression, survival dictates that they relocate. Sometimes migrations are due to forced relocation, as with the forced march of the Navajo people 300 miles from their homelands to Bosque Redondo in 1864, or the Long March of communists organized by the Red Army in China in 1934. The decades since the middle of the twentieth century have seen even more massive movements of people: the forced movement of Muslims and Hindus when India was partitioned in 1947; the relocation of Arabs at the partitioning of Palestine in 1948 and the migration of Jews from Arab countries after the formation of Israel; migrations in flight from ethnic cleansing campaigns in the Balkans, Myanmar, and sub-Saharan African countries; and the millions who have fled Syria's catastrophic civil war.

Whether migration is driven by fear or by a desire to find better economic opportunities, the conditions of migration are usually harsh, and the physical and emotional costs are high. In Europe we see this more than ever in the desperate attempts of migrants to move north from a range of sub-Saharan countries by often illegal means from Libya's thousand-mile coastline. They flee states where war and continued factional conflict have devastated infrastructure and the economy, and poor government and corruption have undermined democracy. Political oppression, forced conscription, and arbitrary imprisonment or enslavement have left entire nations in ruins. But, as Alexander Pope wrote in 1733, "Hope springs eternal in the human breast."[6] Migration offers the hope of finding a better niche in which to live.

Migrants can face considerable hardships raising a family in their new country. Simply getting access to routine medical and social services can be difficult. In the UK this seems to be true even for those who have lived in the country for more than a decade. All the same, recent studies show that migrants tend to be healthier than the general populations of their adopted countries, and for centuries migrants have been invaluable workers and sources of creativity, ingenuity, and cultural diversity.[7]

Finding a Better Niche

Relocation by a species to find an environment in which it can thrive—to find a niche that fits its evolved capacities—is termed *niche matching* or *niche selection*. Members of the species can occupy and thrive in any niche for which their evolved biology makes them adapted. It need not be the environment in which they originally evolved, as the gardener dealing with invasive weeds knows well. A wide range of animals that, through evolutionary processes, were adapted to other environments can thrive in urban environments. City dwellers in many parts of the world know that if they leave their household food waste outdoors overnight in unsecured bins, their garbage might end up scattered about by nocturnal scavengers—

foxes, raccoons, skunks, or even coyotes and bears. Garbage cans are easy prey that don't require the exertions of hunting. Some of these animals will travel quite long distances each night to gain access to the food and warmth of the nighttime niche of the city, returning to hide in less-urban places during the day.

Some of the most dramatic examples of niche selection come from migrant species that occupy more than one niche and alternate between them seasonally. Many such migrations cover long distances and are in themselves feats of survival. Cabbage white butterflies in an English hedgerow on a summer's day seem to be blown randomly from blossom to blossom by the slightest breeze, but many have migrated more than 500 kilometers from south to north and vice versa to follow the seasonal changes in climate. Wind helps many of their bigger cousins, the monarch butterflies, which have a wingspan of ten centimeters, make their flight across the Gulf of Mexico from the northern United States and southern Canada at the end of summer.[8] The monarchs' migration covers 4,800 kilometers, and the total trip is made, not by any individual butterfly generation, but by several generations of butterfly. Many bird and whale species also have long-distance seasonal migration cycles.

Why do some species' ways of surviving, of selecting a niche, involve such challenging feats of migration? Answering this question requires understanding the evolutionary path the species has taken in relation to its selected niche. The species and the niche are in a way inseparable, as has recently been shown for the migration of the red-necked phalarope.[9] These small shorebirds, which probably evolved in the Arctic Circle north of Newfoundland, have gradually occupied areas farther east, even spending summers in the remote islands off the west coast of Scotland. It used to be believed that all these birds migrated southeast to winter on the Arabian Peninsula. Once it became possible to fit tracking devices to some of them, it was found that some phalaropes flew in the opposite direction, across the Bering Strait, down the east coast of Canada and the United States, next crossing to the Pacific coast in Central America, and finally arriving at their summer haunts in Peru. Although the birds were able to stop and rest en route, this journey of 22,000 kilometers is an incredible achievement. It raises the question: Why they would do this, given that the sites on the Arabian Peninsula are so much nearer? The answer appears to be that a population of phalaropes had selected this South American niche by chance but found that it worked for them. Subsequent generations have inherited this adaptation.

Making a Better Niche

Another strategy is *niche construction*—altering the environment to create a relatively constant niche in which the species can thrive.[10] The termites we encountered in the Northern Territory of Australia are an excellent example of this. Their north–south oriented mounds provide a stable internal environment that minimizes the day / night cycle of temperature extremes. Another well-known example is the beaver. Beavers construct their niche by damming a river or stream with logs and branches. They gnaw at trees to fell them and then float them downstream to their construction site, where they build a lodge that on the outside provides a platform where they can view their surroundings in safety from predators on land and that on the inside provides a dry and warm den above the water line. The damming process also creates a pond upstream, which will trap fish and make it easier for young beavers to learn to swim and fish.

The niche that the beaver has constructed is more than just the dam and the lodge. The resultant upstream flooding changes the nature of a much larger area. Plants and animals that occupied the former riverbank may no longer be able to do so, and the range of species in this habitat may change. Apart from the effects on other species, the dam may affect beavers themselves, in terms of food sources, predator numbers, and possibly exposure to pathogens. We can envisage that both aspects of the "extended phenotype" of the beaver—that is, the characteristics of the beaver and also of the physical and biotic features of the niche they have constructed—will be subject to selection pressure and will therefore evolve.[11]

So we can conceptualize the evolution of a niche-constructing species. It starts as an interaction between the characteristics of the species and the characteristics of the niche it inhabits. As for any species, there will be naturally occurring variation in phenotypic characteristics between individuals. Similarly, there will be a range of niches constructed, resulting from differing behavior of the constructors and other aspects of the environment such as its physical characteristics, the level of shelter, availability of food, presence of predators, and weather patterns. Just like for other phenotype–environment interactions, natural selection will favor those individuals best adapted to their found or constructed niche and thus possessing the greatest Darwinian fitness (that is, having the greatest likelihood of successful reproduction and in turn their progeny having the greatest likelihood of survival to reproduce). Inheritance of those characteristics, to the extent they are heritable, by their offspring will lead to a greater preponderance of their phenotypes in the population as a whole. Offspring will then construct niches similar to those of their parents, to which, in the absence of other disturbing factors, they will be well adapted—and so on.

In the absence of some external disturbance or further environmental change, a steady state will be reached between the range of phenotypes in the population and the range of niches they construct. Under these conditions the niche constructed by the population will be relatively stereotyped from one generation to the next. If the environment is disturbed in a significant way—for example, if a drought severely reduces the amount of water flowing down the creek where beavers constructed a dam—the species might have limited ability to respond successfully. Its limited range of evolved physiology and behaviors can be selected only to the extent that there is variation in these behaviors and that they are, at least to some extent, hereditary. Under most circumstances, the degree of phenotypic variation as informed by genetic variation is subtle and thus it takes multiple generations for the population's typical phenotype to change. This process generally will be too slow to meet the challenges of sudden environmental changes, such as the disappearance of the beavers' constructed pond during drought.

The beaver's niche is fundamentally stereotyped because of transgenerational transmission of information. Whether the knowledge and the physical skills necessary to construct its dam and its lodge are genetically programmed into each new generation of beavers, or whether the young beavers learn this from their parents, the effect is the same. The niche they construct is perfect for them—unless the environment changes substantially. The arrival of human fur hunters introduced a new predator to their environment from which the constructed niche could no longer protect them, and so beaver populations fell. And now there are new threats from climate change, which could significantly alter the amount of rainfall in the environments in which they thrive.

For such reasons, biologists have pondered the importance of niche-constructing behavior in a species to evolutionary processes.[12] It can be argued that, precisely because living in a constructed niche makes environmental conditions more stable, this diminishes the part played by environmental variations in evolution of that species. For example, in a species living in a hot, dry environment, characteristics that help survival in that environment will be adaptive; they will favor survival to reproduce and so promote Darwinian fitness. This is what the termite mound does for the termite—it creates a thermally constant environment to which the termite is well adapted. Arguably, then, niche construction buffers the organism against environmental variation and thus creates less "selection pressure."

Niche Modifiers

Humans take the principle of niche construction a step further. We not only construct niches in which to live—in the Arctic, the Sahel, the Himalayas, the Amazon Basin—but we continuously *modify* these niches.[13] We don't simply counteract environmental changes to live in a stereotyped niche—we haven't made a Paleolithic cave as safe and comfortable as possible and held on to this niche for millennia, like beavers have done with their lodges. Instead we use our tremendous ingenuity to develop technological innovations to change our environment, not only to make it better for our survival but also to make it more comfortable, safer, and controllable. We no longer change nature simply to promote our biological fitness: we want to make it nicer to live in too. We never stop using our ingenuity to do so and this has consequences: because through our emerging technologies, whether intentionally or not, we are continuously modifying our environment further.

Currently we are in one of the hottest, driest periods in the past hundred years. Wildfires now rampage in areas where formerly they were almost unheard of—such as the Yorkshire Dales and northern Scandinavia. In New South Wales, livestock have been dying because they have no water. Many animals have perished, and some species may even be threatened with extinction. Not so for humans. It is true that some vulnerable people, especially the elderly, are at risk, and there have been tragic human fatalities due to the changing weather.

But nobody is yet talking about our extinction as a species. We created air conditioning to cool us down; we drill wells to provide ourselves with water, and even truck water in if local supply runs out; we can desalinate seawater if we need to. Satellite data help us predict and prepare for the challenges posed by the weather. We use all the technological tricks we have invented and perfected to modify our niches continuously in order to minimize any threat that they pose.

Initially the development of shelter, the control of fire, the development of spears and hunting weapons and of clothing were behavioral and technological adaptations that extended our physiological capacities and allowed us to spread over the planet and thrive as a generalist species in many very different environments. Our Darwinian fitness—the survival of each generation to reproductive age in sufficient numbers to maintain or increase the population size—was well served by our Paleolithic niches. But our species has progressively built technological enhancement upon technological enhancement in a way no other species has done. In distinction to any other species, humans have a set of cultural drivers underpinned by our evolved biology to achieve a broader range of social and societal goals. With few exceptions—such as, perhaps, the isolated Amazonian groups who have not had contact with modern societies—our social structures have changed dramatically over the past 10,000 years in parallel with our suite of technologies. Human technologies did not stop developing just because survival was better assured by niche construction, or simply because inhabiting a cave gave protection against the weather and predators. Our ingenuity drove us to cumulatively modify our niches and, in doing so, created the complex physical and social milieus found in modern society. Every aspect of the human condition—how we feed ourselves, how we engage with others, how we spend our time, how we confront illness, how we behave—has been affected by continued technological modification. We have modified our environment beyond what we require to survive.

Living with the Enemy

Our environment also includes our relationship with plants, animals, and particularly microorganisms such as viruses, bacteria, and parasites. As the human genome has been explored, we have found in our DNA evidence of our ancestors' interactions with the environment and in some cases their success in overcoming the challenges it has posed. For example, our DNA contains considerable evidence of past viral invasion.[14] Even our cells' mitochondria, which are the intracellular organelles that effectively use oxygen in the generation of energy, and their plant equivalent, the chloroplasts, have been shown to originate as bacteria that were captured by ancient single-celled organisms and co-opted to provide energy-generating machinery.[15]

Our evolution did not occur in an abiotic environment. It has been normal and inevitable to be exposed to many microorganisms, and to a certain extent our immune systems have been tuned to meet the challenge. They may even react in inappropriate ways if they do not meet such a challenge. Our modern focus on hygiene, eliminating worms and other pathogens from our guts and respiratory systems, has played a role in the rising incidence of asthma and some autoimmune diseases—so much so that one therapy now being used for Crohn's disease is the use of extracts derived from such helminthic pathogens.[16]

Indeed, the shaping of the human genome in certain parts of the world has been very much influenced by patterns of infectious disease. In areas where malaria is common, individuals with mutations in the molecule hemoglobin are more likely to survive. This is because the malaria parasite normally lives within red blood cells, but the mutation changes the biology of these cells and they are less likely to be infected by the parasite. We can see how such a mutation would have spread through the population and been preserved through providing a survival advantage against the malarial parasite. However, like almost all of our genes, we inherit two copies of it—one from each parent. Inheriting a single copy of the mutated gene gives relative protection against malaria, but in persons who have inherited two copies of the mutation the red blood cells become particularly distorted from their normal disc shape to elongated sickle-shaped cells. The resulting sickle cell disease is very debilitating. Here selection pressures have been balanced between favoring those with one copy of the mutated gene, who are likely to live longer because malaria can be fatal, and those with two copies, who will develop sickle cell disease. This is not an isolated case—there are other mutations that also cause malarial resistance when only one copy of the gene is present and a disease when two copies are present.[17]

The pace of technological discovery has been rising—in many ways we can see the last 200 years, and particularly the last 50 years, as a period of exponentially rapid increase in collective knowledge. The scientific method has been consolidated, assisted by the emergence of many technologies which in turn allowed science to progress even further and faster. But in reality, our understanding of physiology and medicine is very young. The medicine of Hippocrates and Galen did not include microorganisms, and our understanding of the gene is less than a hundred years old. The first antibiotic is even younger.

Dealing with pathogens remains the fundamental basis of many public health campaigns, including promoting hygiene, use of antibacterial agents, immunization, and isolating pockets of disease such as Ebola to prevent its spread. From the discovery of "germs" and the first programs of hygiene in the mid-1800s, public health measures have reduced mortality and virtually eliminated many infectious diseases, such as smallpox and polio.

But, as we will continue to explore in later chapters, interfering with nature often has a price. Our own biological evolution might no longer be of importance in confronting pathogens because of the weapons we now have to fight them, but pathogens' continuing evolution definitely is important: microorganisms have fast reproduction times and can therefore quickly evolve to be resistant to our weapons. For instance, every time a new antibiotic is discovered and put to use, strains of bacteria soon appear that are resistant to that antibiotic. There is an increasing fear that because of our overuse of antibiotics, we will soon face again bacterial scourges that we thought we had eradicated decades ago.

Our biotic environment is not just outside of us, it is also within us. Our bodies harbor—on our skin, in our respiratory systems, and even more in our intestines—an enormous number of viruses and bacteria of many different species. In fact, there are far more bacterial cells in or on our bodies than the cells we consider to be "us." Most of these bacteria are harmless, and many are beneficial to our lives. It appears that over time we have coevolved with these beneficial organisms in a symbiotic relationship. We provide an environment that suits them and a source of nutrition, and they in turn provide useful protective roles by assisting in digestion, immune function, appetite control, and many other processes.

The peaceful coexistence, or "symbiosis," between ourselves and our microbiotic environment raises an interesting, albeit somewhat philosophical, question: We are not healthy without the appropriate colonies of bacteria inside our intestines, so where is the boundary between us and our external environment? Are our gut bacteria part of us or not? At first sight, this seems to be a new question, but in reality it may not be. The beaver is not healthy without its lodge, the termite is not healthy without its mound, and the tunnel spider is not healthy without its web tunnel. So the boundary between them and their environments is not absolute. This led Richard Dawkins to coin the term "extended phenotype" in his book of the same name, to refer to those characteristics of the organism that, although external to its organic boundaries, are integral components of its core viability and characteristics.[18]

Up to this point we have been discussing Darwinian fitness largely in terms of survival until having the opportunity to reproduce. But for sexually reproducing organisms—all birds and mammals, and most fish, reptiles, amphibia, and insects—that opportunity requires finding a mate, and in a population of individuals this involves some element of "choice," whether conscious or not.[19] This problem challenged Darwin and the other architect of evolutionary thinking in Victorian England, Alfred Russel Wallace. They debated the nature of "sexual selection," and the associated issues of human emotions and consciousness, for many years, and when Darwin died, in 1882, they still had not resolved their differences of opinion.

Sexual Selection

Alfred Russel Wallace was a collector, a self-taught scholar, and, like Darwin, a naturalist. In 1855 he wrote a very important paper, building on the ideas of Alexander von Humboldt, linking species distribution to geography and pointing out that species living in similar regions appear to have similar characteristics. Although this paper was well known to Darwin, who had first written his notes on natural selection for his own records in 1842, and even though Darwin and Wallace were active correspondents, Darwin failed to recognize how close Wallace's emergent ideas on the origin of species were to his own. When Wallace was on the Island of Ternate in the Indonesian archipelago in 1858, his ideas gelled and he formulated his own concept of natural selection as the mechanism to explain speciation.[20] He sent the draft of a paper to Darwin. Darwin was perhaps panicked by this, but it motivated him to publicize his own well-developed ideas and he was able to claim precedence as a result of his much earlier note. Darwin's friends Lyell and Joseph Hooker arranged for both his note and Wallace's to be presented one after the other at the Linnaean Society (a natural history club) in London in 1858. Even though Darwin had been working for years to produce a long volume on natural selection, he rushed to produce a shorter one. *On the Origin of Species by Means of Natural Selection* was published within a year and very quickly became widely read. The longer version was never completed, although he was to write many other books which encompassed that material.

For the rest of their lives, Darwin and Wallace debated the function and evolutionary significance of the dramatic sexual-display ornaments and behaviors of various species. The best-known example is the peacock's tail, with its striking iridescent blues and greens and characteristic "eye" patterns. In contrast, the peahen's coloration is very muted and mostly a pale brown. In his travels in Southeast Asia, Wallace had come across (and shot and collected) many other examples of such "sexual dimorphism." The male superb bird-of-paradise is able to raise the display feathers on his neck into a large black fan above his head, with a turquoise gash and two little white spots that give it the look of a carnival mask. The female, in contrast, is a rather drab, reddish-brown bird with a black head. Many other male birds engage in an elaborate dance performance involving acrobatic feats and stylized courtship gestures. What could be the functions of these extraordinary ornaments, displays, and behavior? Presumably, if they exist in so many species in so many different forms, they must have some functional importance. If they are products of evolution, they must convey some advantage in terms of species survival and reproduction.

While Darwin was finishing *The Origin of Species,* he was developing the concept that, along with natural selection, an additional form of selection operated in many species: sexual selection. But he dared not write about that in *The Origin of Species,* which was ready to be sent to press. He was aware that much criticism already awaited him, for his challenging the biblical idea of divine creation of the world and its life-forms, without adding sex to the book. He was a cautious man: on one hand, an innovative and iconoclastic thinker; on the other hand, anxious to remain an accepted member of the landed gentry and the establishment. That caution extended to his science. He took many years to refine his concepts, particularly of sexual selection. It was not until 1871 that he addressed the issue in his other major work, *The Descent of Man, and Selection in Relation to Sex.* Here he dared to broach his idea of sexual selection in terms of both male-on-male conflict and mate choice, adding that in most species with clear differences in appearance between males and females, it was the female who made the choice.

Darwin wrote about birds whose males gather at one spot, called a *lek,* to engage in display competitions in front of a female. Male ruffs, for example, compete with erectile displays of their ruff feathers. The one that most impresses the female is the one that gets to mate with her. Darwin suggested, regarding the peacock's tail, that it might be the beauty of the display that impressed the peahen.

Wallace held a different view of sexual dimorphism, the distinctive appearance of males and females. He thought it had to be purely adaptive—in other words, that it must confer a direct advantage to the successful mating pair in terms of their production of numbers of healthy offspring. He was unable to accept Darwin's view that animals other than humans could have emotions or any concept of beauty or elegance, and therefore did not agree that mate choice might be based on such concepts. For Wallace the sexual display by the male had to indicate to the female his characteristics that would contribute to healthy offspring, such as his nutrition, his strength, or even his parentage. Perhaps the heavy tail the peacock carried around, which was of no use for anything other than display, was in fact a measure of his strength. The competition for the peahen would therefore come down to males' ability to grow and carry an extravagant tail. This might be seen as a more aesthetically pleasing example of the principle of male competition through physical violence, as in the rutting-season combat between male deer, who battle with their antlers to win the right to mate with females. The defeated males, often injured, are excluded from the herd and are more vulnerable to predators. Similarly, male elephant seals rely on their immense body size to jealously guard their harems on the beaches of South Georgia, and they will engage in vicious fights against intruders.

Today most evolutionary biologists believe that there is truth in both Darwin's and Wallace's ideas, and that the relative importance may vary between species. The male bower bird of Australia, for example, spends much time building his bower, of a particular shape according to the species, as a display to attract the female. The bowers are elaborately decorated with shells, stones, and leaves—and these days, sadly, with colorful bottle tops and pieces of plastic. The selection can be highly specific to a particular color. This does not seem to have much to do with the male's strength or his health as a potential father. Adorning the mating site is also a strategy for short-tailed bats, one of only two species of indigenous terrestrial mammal in New Zealand (the other is also a bat). They prepare special mating sites in the holes of trees and urinate over themselves and in the nests themselves to make them more attractive to females.[21] While most examples of mate choice involve males competing for the attention of females, there are some animal examples of female ornamentation evolving for male attention.[22]

Sexual selection can occur either through mate choice or through competition between males for mating rights. In many species the contest is simply played out in terms of body size, where males are bigger than females and the largest males are more likely to win in the mating game. Other species, such as lions, will fight to death or exclusion for dominance and possession of a pride. Some evolutionary psychologists have proposed that male-on-male conflict underpins the development of weapons by humans.

There is an implication in the idea of sexual selection—whether by mate choice or sexual conflict—that such biology can speed up the process of selection. The great twentieth-century mathematical biologist Ronald Fisher termed this "runaway" selection.[23] If having bigger antlers makes success in the mating game more likely, then over time the genes controlling antler growth would be favored, and antler size would continue to increase. Or peacocks' tails would become larger and brighter. And this is indeed what seems to happen. Records show that the Irish or European elk grew antlers that were more than 3.5 meters in width and weighed over 40 kilograms.[24] But this process could not go on forever—the positive feedback will drive it to an intolerable point, just as such feedback does in a sound amplification system. At some point huge antlers would become a handicap—limiting the animal's ability to move through a forest or escape predators. A male African elephant could not be very much larger without enormously thick bones to hold its weight—and even at its present size it usually sleeps standing up because lying down risks fracturing its ribs.

Darwin's idea implies that the mechanisms underlying the female's aesthetic appreciation of "beauty" in a species would have to evolve simultaneously with the "beautiful" sexual ornaments of the male, and both would have to vary across a population if selection was to be useful. Indeed, Darwin was fascinated by the concept that animals might have a sense of beauty and undertook many experiments at Down House trying to prove it. Equally, Wallace's argument resurfaced in an Israeli biologist's concept of the handicapping principle. Amotz Zahavi argued that the peacock's tail or the lion's mane or the bower-bird's enormous efforts to amass his collection were all ways of the male showing that he was strong and healthy enough to indulge in wasteful or energy-consuming activities that might handicap him from a survival point of view.[25] From the female's perspective, selecting a mate who had such evidence of strength and health would advantage her potential offspring.

The Fashion Industry

Darwin did not want to get into a discussion of sexual selection in humans, to avoid even more controversy over his ideas. But while much of the argument in *The Descent of Man* was about birds—species in which the female makes the selection of a mate—in humans it was usually the male who was responsible for the selection. This does not mean that it is only the female who has to appear attractive. Probably from the very earliest time when our hominid ancestors started to wear simple clothes, these garments took on a significance beyond just protection from the elements or modesty. The type of animal skin or plant fibers used to make clothes could indicate status as well as gender. Soon other forms of decoration were developed, from body decoration to beads of bone or colored stone.

We cannot put a date on the evolution of the fashion industry—the crafting and trade of adornments. Archaeological discoveries have found artifacts that may have been ornaments dating back at least

120,000 years, and Neanderthals may have had a sense of ornamentation, although this may have been copied from their human cousins.[26] For us, clothes and accessories have always had a function beyond just sexual selection, conveying social signals and also supporting self-esteem—and Darwin had more than a passing interest in female fashions in clothing.[27] From ancient times humans have been willing to undergo severe discomfort to make their bodies conform to socially accepted norms, from the huge discs in the earlobes of some Ethiopian tribes or the neck-stretching rings of some African and Asian tribes to the head compression from birth in some American Indian tribes or the binding of feet in China. We might think, too, of the massive industry of cosmetic surgery today, or the widespread use of hormone replacement therapy to prevent the visible signs of aging. These technologies have costs for our health as well as benefits for our self-esteem. Our ingenuity enables us to change not only our world but our bodies, too.

In the Light of Evolution

Our capacities for progressive and cumulative niche modification stand out from those of other species not only in degree but also in their consequences. But this does not mean that humans have a distinct evolutionary origin—a debate that we have already seen involving comparisons of gorillas to humans. The question of human origins was also one that Wallace and Darwin debated at length. Wallace argued that, because of the distinctiveness (in his view) of human consciousness and our range of emotions, there must have been some form of special evolution of humans. This concept Darwin rejected, preferring to focus on the continuity between humans and the rest of life. Darwin's view has prevailed, but the sheer scale of the seemingly special human characteristics—highly developed cooperation, altruism, and morality, along with our technological ingenuity—continues to be the focus of debate among evolutionary biologists and philosophers.

The Russian émigré evolutionary biologist Theodore Dobzhansky said that "nothing makes sense in biology except in the light of evolution."[28] What he meant is that evolutionary biology is *the* integrating principle of all aspects of life science, from the most reductive molecular biology of tiny components of cells to the exploration of the psychology of human behavior in crowds. Evolution is the consequence of the game of survival playing out over many generations through a series of interactions between our genes, our development, and our environment. As humans we use our ingenuity to change nature, to continually modify the niches we inhabit, and in part to win the game of survival by reducing the effects of natural selection upon the variety of members of our species. We use great ingenuity to exploit the processes of mate selection, although this is far from simple in our species, involving complexities that go well beyond our biology. But as the concept of niche modification illustrates, we use our ingenuity for non-adaptive motives.

Fundamental to the concept of evolution is the passage of genetic information to the next generation. This is usually thought of simply in terms of our genes. But is this all there is to inheritance? In Chapter 3 we explore the limitations of this idea.

3 | INHERITANCE

THE PACE OF DISCOVERY in fundamental biology rose exponentially during the twentieth century. Gregor Mendel's work on the particulate basis of inheritance was not rediscovered until early in that century.[1] This led to the concept of the gene and its chemical nature in DNA found in long sequences called chromosomes, which in turn led to new understanding of the processes of inheritance and variation essential to understanding evolution. For a while, evolutionary concepts were marginalized. The challenge was how to combine the concepts of the gene and those of evolutionary biology. The Mendelian model of the gene implied inheritance of discrete characteristics, whereas natural selection was based on continuous variation in the phenotype. How were these to be reconciled? The solution was found in the application of mathematics to population genetics. In the 1930s a group of very influential population geneticists, biologists, and theorists, the so-called neo-Darwinists, recognized their compatibility: they produced what became known as the "Modern Synthesis," and until very recently this concept has dominated biological thinking about what makes us what we are. Their view was that evolution could be reduced to a shift in the frequency of different gene variants in a population over time. It was really only in the opening years of the twenty-first century that we began to see that this view is too narrow—it excludes important concepts about the ongoing roles of the environment and development, and the possibility of forms of inheritance other than by genes alone. We will explore these concepts in this chapter.

Darwin had mused over a question that was to engage neo-Darwinists in increasingly heated debate, which has continued until the present. Even though Darwin largely considered selection in terms of whether an individual of a species was fit to survive and reproduce, he also thought that selection could exist at the level of a group as a result of its collective characteristics and behavior. Behaviors within one group might favor its survival over another group with different practices. The potential for such *group selection* to explain different behaviors was attractive. But in the 1960s the concept had fallen out of favor, largely in response to a teleological argument that was derived from an extreme, and now totally discounted, interpretation of group selection—namely, that selection operated at the group level for the "good of the species."[2] The response to this latter argument was robust and rapid, led by the American evolutionary biologist George Williams, who emphasized that selection was about the individual, not the species.[3] Game theorists such as John Maynard Smith added to the attack on group selection by highlighting the issue of the freeloader: When a group includes a person who does not abide by its rules, she or he could gain advantage at the expense of others in the group, much as a criminal gains at the expense of law-abiding citizens.[4] Modeling led the game theorists to conclude that group selection must inevitably fail: There would be no reason for any individual to be law-abiding, and so the group would soon become dominated by freeloaders. Hence, such a group could not survive.

And then the debate went a step further. In *The Selfish Gene* (1976) Richard Dawkins argued that our individual bodies are no more than vehicles transporting our genes and responding to the genes' need to perpetuate themselves.[5] That there might be higher levels of selection—involving groups, for example—or that passage of genetic material to the next generation was not the only goal of life, was seen as impossible, if not heretical, in this new religion of the gene. This strong view of genetic determinism was reinforced by the growth of genetic technologies. Hyperbole about the role of the gene became the norm as enormous sums were invested in genomic science, especially the Human Genome Project.[6] Such science has undoubtedly led to enormous increases in our understanding of biological processes, but the more we learn from it, the more we recognize that there are many more layers of complexity to what makes us what we are than just the genes we inherited from our parents.

There has emerged a much more nuanced, but still somewhat contested, view that selection can operate at multiple levels. Some new insights have come from thinking about biological systems that operate below the level of the individual: for example, the biology of cancer can involve selection of an individual cell which has mutations that make it more likely to divide, leading to the growth of a clone of cells in a tumor that expands at the expense of other cells surrounding it. And while selection at the level of the organism is the dominant view, group selection is no longer dismissed as impossible.

The term *multilevel selection* has been introduced to indicate the generally accepted view that selection generally occurs at an individual level but that, under some situations, group-level selection can occur and could well have played some role in how humans evolved to have particular social behaviors.[7] A further refinement, which we will discuss in Chapter 4, is known as *cultural selection*—selection determined by cultural behaviors of the group rather than biological traits embedded in genes. Debate continues over how important group selection is, aside from in organisms that have very complex genetic interrelationships such as social insects like colony-forming bees, wasps, termites, and ants.[8] We will return to this in Chapter 4 when we consider the role of culture in human evolution.

Beyond the Gene

While the very gene-centric, deterministic view of life was reinforced by the Human Genome Project, ironically the Human Genome Project also served to confuse the definition of a gene.[9] No longer could a gene be seen as simply a sequence of nucleotides in DNA that coded for RNA, which in turn coded for a protein—the formulation that became known as the "central dogma" of molecular biology, first proposed in 1958 by the co-discoverer of the double helix structure of DNA, Francis Crick.[10] Rather than the information flow operating only in one direction, as Crick had suggested, we now know that proteins and RNA can bind to DNA to change its function. Moreover, a single gene, which is one segment of DNA, can be involved in making multiple proteins. The amount of DNA in our cells is substantial, but only a small fraction of it can be labeled as "genes"—that is, as sequences coding for proteins (we have about 22,000 genes). This discovery was made possible only by the development of technologies that could map the sequences of the four nucleotides that constitute the alphabet of DNA. Most of the rest of the DNA was first thought to be "junk," left over from previous phases of evolution, genetic material inserted by viruses, and so on. Now we know that a much larger fraction of the nucleotide chain in human DNA appears to be capable of functioning, even if its function is not to produce proteins directly but to regulate the function of the coding genes. Far from simplifying life by providing a blueprint or a map, the Human Genome Project has revealed whole new levels of biological complexity.[11]

The rush to sequence the human genome was founded, and funded, in no small part on the genetic determinist view that once we knew a person's genetic identity, his or her destiny could be foretold. Billions of dollars, pounds, and euros later, we know it is not that simple. Genomics has assisted medical research enormously and has started to have some value in diagnosis and in choice of therapy, especially in cancer. But at the same time there has developed a large and highly profitable industry in genetic measurement, most of which has very little scientific validity but charges us handsomely for genetic analysis that purports to tell us who we should mate with, what we should eat, what diseases would be most likely to affect us, who our ancestors were, and much more. Extraordinary and misleading claims abound, both within this industry and in the popular media, about the certainty that genetic information provides.[12]

The genetic component of many common diseases is smaller than often claimed and involves many genes and new layers of complexity—particularly those involving development and our responses to the environment. Events in early life leave echoes in our bodies and minds for the rest of our lives. As we will see later in this chapter, some of these are through the mechanisms of developmental plasticity and epigenetic processes, and some are through embedded memories and learning in our immune systems and brains—the latter involving changes that are reflected in altered connections and development of brain areas.

Genetic science is extraordinarily valuable, but the information it provides is only one part of a much bigger story. And while a more nuanced understanding of multiple layers of biological information is now emerging, there remains intense debate about how to integrate these broader insights with the fundamentally well-understood processes of genetic inheritance, natural selection, and evolution. It is possible that gene-centric dogmatism arose partly as a defense against both the theistic and the erroneous proto-Lamarckian models of evolution that emerged in the decades after Darwin's great insights. It is in the nature of science, though, that intellectual frameworks must evolve as new knowledge is generated and integrated. Despite the ongoing and ultimately constructive debates within the evolutionary biology community, it is nonetheless remarkable that the general framework that Darwin proposed, at a time when the molecular basis of inheritance was not understood, remains firmly at the core of biological understanding even today. We need now to discuss newer ideas about inheritance. But as is so often the case in evolutionary biology, when we discover something new we also find that Darwin himself had indicated that it might exist, even though the science of his day was not able to uncover it.[13]

Inheriting the Environment

Despite their disagreements about sexual selection and human emotions, Wallace and Darwin did share a perspective on evolution that accords with the views of virtually all evolutionary biologists even today: that the driving force for evolution is survival to allow for reproductive success. Obviously, without reproduction there can be no passage of genetic material from one generation to the next. However, our lives are influenced by far more than inherited genetic information.

When conditions are stable, aspects of the environment affecting survival to reproduce will be experienced by multiple generations. But could information about the environment itself be *passed* from one generation to the next by becoming embodied in some way? Some evolutionary biologists argue that because, for humans, our characteristics determine the environment we create, the niche we modify, and because those characteristics are ultimately determined in turn by our genes, it follows that culture and human inheritance, particularly genetic inheritance, are inseparably intertwined. This is the concept of gene–culture coevolution, which we will return to in Chapter 4. Right now, we need to think about a further possibility—"non-genomic" but biological inheritance. If we discover that such mechanisms exist, it strengthens the need to consider a broader range of evolutionary mechanisms.

The idea that the environment in one generation might, by altering the characteristics of adult individuals, impact on the next generation is not new. It is usually associated with the notion of the inheritance of acquired characteristics, an idea that is often ascribed to Lamarck but was broadly accepted at that time—even though, in its crudest form, it can easily be shown to be implausible. A female dog losing a leg in an accident is not likely to give birth to a three-legged puppy. And any sheep farmer who has been docking lambs' tails over generations of his flock to stop flies from breeding in the dung that clings to their tails, knows that inheritance of acquired characteristics at that level does not occur—new lambs need to have their tails docked. Jewish boys still are born with a foreskin despite 4,000 years and perhaps 200 generations of circumcision.

August Weismann was a great German biologist and an enthusiastic supporter of Darwin. He saw these obvious flaws in crude Lamarckism, and to make the point he conducted an experiment in which he cut off the tails of mice over many generations, in order to see whether their offspring would also be tail-less.[14] Needless to say, they were not. His point was that damage to an animal produced by surgical means could in no way be seen to produce an adaptive phenotype: the mice were not better able to survive and to breed in their cages in the lab as a result losing their tails. Darwin himself, though, did not reject or exclude the possibility of such "use-inheritance," and he made enquiries to see whether the children of blacksmiths had more muscular arms than other children.

But more subtle experiments are able to confirm the possibility of what we now call "non-genomic" inheritance at least over a limited number of generations. For example, in the 1960s Tracy Sonneborn, working with the single-celled organism *Paramecium,* surgically altered the orientation of cilia—hair-like projections that beat in a beautiful rhythm, enabling unicellular organisms to swim. Sonneborn surgically removed a section of the cell wall, turned it 180 degrees, and replaced it on the cell. The section continued to beat as it did before removal, but now in the opposite direction to the other cilia of the *Paramecium.* This peculiarity was inherited by the two daughter cells of the organism, and to a lesser extent even by the granddaughters.[15] Because the genetic information in the cells had not changed, Sonneborn's work is often credited as a good example of non-genomic inheritance.

Lamarck Redux

There was another wave of interest in neo-Lamarckism early in the twentieth century. Paul Kammerer, an Austrian biologist who studied amphibians, worked in a research institute in Vienna during the first two decades of the twentieth century, where he started manipulating the breeding conditions of salamanders and toads.[16] The species of salamander he studied either bred in wet conditions, spawning tadpoles, or in dry conditions, giving birth to fully formed salamanders. He claimed that he could induce salamanders that were initially bred in dry conditions to give birth to the adult form over several generations even after they were returned to wet conditions. This, he suggested, was evidence of inheritance of acquired characteristics—in the form of a "memory" of their ancestors' dry conditions. The work was highly acclaimed and he received several prizes for it.

Kammerer also studied a species of toad, the midwife toad, that breeds on land. During mating the male mounts the female and holds on to her rough skin. Some of the males' feet have swellings, so-called nuptial pads, that enable them to hold on to the female's slippery skin even in wet conditions. Kammerer managed to breed midwife toads that did not have nuptial pads in wet conditions, and he claimed that under these conditions the male offspring developed these pads *de novo;* even more surprising was his claim that the pads were found in subsequent generations of toads even if they bred in dry conditions. This was clear evidence that induced characteristics could be inherited over several generations—which seemed like strong evidence for Lamarck's ideas.

The trouble was that other biologists could not repeat Kammerer's experiments. The famous evolutionary biologist William Bateson was particularly skeptical.[17] Unfortunately, Kammerer's work was interrupted by the First World War and many of his animals and specimens were lost. But after the war the debates continued—Kammerer claimed he was right, and suggested that his studies in a very different species, the sea squirt, also gave evidence of the inheritance of acquired characteristics. But others could not repeat these experimental results either.

Kammerer transiently moved to the Soviet Union, since his ideas fitted very well with those of the influential Soviet pseudo-scientist Trofim Lysenko, who similarly claimed that environmental conditions could produce heritable effects on the characteristics of plants.[18] Lysenko was strongly supported by Josef Stalin, who was keen to find new ways of increasing crop yields or even influencing human nature. The plot thickens further, though, because Lysenko's work was also widely discredited, and it is now generally considered that he was a charlatan.

When the influential American naturalist Gadwyn Noble inspected a preserved specimen of Kammerer's midwife toad, he found Indian ink in the toad's pads, suggesting that the specimen had been tampered with to enhance the appearance of nuptial pads. Noble started a correspondence with the director of the Austrian Research Institute, saying that he was very suspicious of Kammerer's results. The director supported Kammerer's claims, as did others, but Noble was not persuaded and in 1928 he published a letter in the influential journal *Nature,* saying he believed Kammerer's results were fraudulent. A few days later Kammerer committed suicide. Whether this was related to the allegations of fraud is unclear.

Interest in Kammerer's observations, with all the murkiness surrounding the man, the politics, the science, and the times, has never totally evaporated. Recently some Chilean scientists looked again at his data and at details in the statistics that had not really been considered previously, and suggested that his results could be compatible with recent concepts in epigenetics.[19] However, no one has actually repeated Kammerer's experiments.

The Epigenetic Landscape

In the 1960s a very distinguished British scientist literally shifted the landscape. Conrad Waddington was trained as a developmental biologist and geneticist. Although he was well linked to the mainstream neo-Darwinist community, he was seen as rather a maverick.[20] Critical to the development of his ideas was time he spent in New York in the laboratory of the Nobel Prize winner Thomas Hunt Morgan. Morgan had shown that it was possible to induce genetic mutations in the fruit fly via exposure to ionizing radiation. He showed that these mutations occurred in the chromosomes, and so for the first time firmly linked these structures with the mechanisms of inheritance. We now know that the chromosomes consist of DNA coiled around proteins called histones and that along the DNA are located the genes: the sequences of nucleotides that code for the production of particular proteins.

Like Morgan, Waddington believed that new ideas about genetics could be established through the study of fruit flies. Fruit flies have very clear characteristics that can be seen under the microscope, such as the number of hairs, types of veins in their wings, eye structure, and so on, and these are clearly heritable. Morgan had started by disrupting chromosomes and showing that this led to altered characteristics.

But it was also clear that environmental factors acting in development could influence the organism's characteristics. It had been known for a long time that even organisms with the same genetic makeup could vary in their characteristics or phenotype. It was also known that there were causes of variation in traits that appeared not to be genetic—some characteristics did not "breed true" as would be expected for a purely genetic effect. Waddington became interested in this interplay between genetic variation and developmental variation induced by the environment. Under what circumstances might the environment affect gene function to result in a change in developed characteristics? He recognized that there must be mechanisms that buffered the developing organism to some extent against genetic and indeed environmental variation: he called the process "canalization," a concept that still holds true. But clearly, under some situations mutations or environmental factors could shift the pattern of development, perhaps by revealing cryptic genetic variation.

It sometimes happens in science that our understanding is advanced by someone producing a revealing diagram. Waddington pictured development as a ball running down a valley that bifurcated again and again, with the walls of the valley making the ball follow one path rather than another at each fork in the path. If the sides of the valley were not very steep, the canalization would be weaker, and the path taken by the ball would be less strongly determined. His insight was that the ball might represent a mass of undifferentiated stem cells, and the forks in the path represent points at which successive stages of differentiation during development took place and at which the cells became committed to a certain lineage—so that, by the end of the journey, the cells might have become liver, pancreas, heart, gut cells, and so on. This now-famous picture is called the "epigenetic landscape." The celebrated artist John Piper produced the image on which the first version of Waddington's epigenetic landscape was based.[21]

One of the reasons Waddington's diagram has become so influential is that it elegantly conveys the concept denoted by the term *epigenetics.* Waddington coined the term in about 1940 to mean "around or beyond the gene," although at that time there was no knowledge of the biochemical mechanisms involved. Waddington felt that *epigenetic* was preferable to the term *development,* which was open to different interpretations.

Some of the clearest examples of Waddington's concept of epigenetics come from insects. The female honeybee larva will develop either into a queen bee or, more often, into a worker bee. These forms are both female and have the same genome but are very different in shape, behavior, and biology. Queens lay eggs, workers have inactive ovaries; queens are bigger and their metabolism is tuned for just one flight—the nuptial flight. Workers are smaller, but their metabolism is tuned for frequent flights; workers have mouthparts suited for collecting nectar, queens for fighting other queens to the death. A larva turns into a queen only if it is exclusively fed royal jelly and no other food in its first developmental stages. If it does not get enough of this nutritious food, but instead only nectar, it will develop into a worker bee.[22] This is an example of a simple environmental stimulus acting at one critical window in early development to change multiple components of the mature phenotype.

A particular environmental change that Waddington studied was exposure of fruit flies and their larvae to heat. He found that some of the flies that emerged had a different vein pattern in their wings. If he selected these flies and again exposed them to heat, they produced more offspring with this altered vein pattern, and so on until almost all offspring had the new pattern. After a time, flies with this pattern emerged without their larvae or mothers having been exposed to heat. Waddington called this phenomenon genetic assimilation. Had he done the equivalent of Kammerer's experiment with the midwife toad and induced the inheritance of an environmentally acquired characteristic?

The interpretation of Waddington's experiment is still debated today.[23] Had Waddington induced a new heritable change, presumably in the DNA, by heat shock? Or was it simply that the heat had revealed some "cryptic" variations in the fruit fly genome which were exposed only when the heat reduced canalization? Waddington himself favored the latter explanation. But although this was perhaps most compatible with contemporary genetic concepts, the experiment gave enormous impetus to studying the impact of the environment during development.

Changing the Message

In the double helix of DNA there are only four nucleotide bases, the sugar-amino acid molecules that make up the DNA alphabet. In a protein-coding gene, a triplet of these bases makes up one part of the "message" that informs the making of RNA; this triplet of information turns into the message the RNA uses to code for a specific amino acid; proteins are then made of strings of these amino acids. In 1975 a British geneticist, Malcolm Holliday, showed that one type of these bases could be chemically modified by the attachment of a very simple molecule consisting of one carbon and three hydrogen atoms, a process called methylation.[24] He proposed that this gave another mechanism of control of "gene expression" because methylation in regulatory regions of the DNA was shown to be associated with gene inactivation. DNA is normally tightly coiled around proteins called histones in the chromosome—it has to be coiled up because cells contain about two meters of DNA if it could be pulled into a straight line. Simple chemical changes such as methylation can alter how tightly it is coiled, and how easily the gene code can be read by the ribosomal machinery, giving yet another potential regulatory process.

Understanding these processes was the start of molecular epigenetics. The science did not progress rapidly for another twenty years, until it became apparent that epigenetic modifications by DNA methylation and changes in histone chemistry could explain another important paradox: We are formed from one fertilized egg, which means that the DNA in all our cells is the same (except for cancerous cells, which generally have DNA mutations). But some of the cells of the embryo turn into skin cells, some into brain cells, some into muscle cells, and so forth. And when we look at the pattern of gene expression—the genes that are active within these different cell types—we find that they are very different. A skin cell will make keratin, a neuron will not. A pancreatic islet cell will make insulin, a heart muscle cell will not. We now know that it is epigenetic differences that turn genes on or off in these different cell types to make them what they are. Much of modern developmental biology aims to understand how cells in early development "know" which epigenetic changes to make—they do so by sensing very subtle chemical differences in their cellular neighborhood. This story was implicit in Waddington's epigenetic landscape.

The utility of Waddington's concept is that it allows for the incorporation of new ideas. As he presented it in the third version of his model, the "landscape" is not fixed like the hills around Edinburgh where Waddington spent much of his working life. Instead, he models it like the canvas of a large tent, held up by poles and kept in place by guy-ropes attached to pegs driven into the ground. These pegs are the only fixed part of the model—like the genes. If we tighten or loosen one guy-rope, this changes all the contours of the "landscape" modeled by the canvas. Some valleys become steeper, others shallower. So the degree of canalization changes from valley to valley. The ability to show the widespread impact of subtle epigenetic processes so simply in this way is the beauty of Waddington's model.[25]

Epigenetic processes are now defined in terms of their impact on how DNA operates to turn genes on or off, and there are many types of epigenetic processes and functions. For example, females have two X chromosomes, but one of these is almost entirely silenced by epigenetic processes that keep its DNA tightly wrapped so its genes cannot be turned on. We have already described how epigenetic processes allow one fertilized egg to have daughter cells that in turn become quite distinct as skin cells or liver cells. But of particular interest to us is the role of epigenetics in development and thus in passing information from one generation to another. This can happen in at least two ways: Environmental signals such as changes in nutrition can alter the developing embryo's epigenetic state and thus gene action and thus the trajectory of development; or there can be epigenetic changes in the sperm or egg, which would be passed to the next generation. In these two ways epigenetic processes can affect the development of the next generation from the embryonic stage onward.

This is often thought of as a present-day return to the ideas of Lamarck, a sort of neo-Lamarckism often rather loosely characterized as the "inheritance of acquired characteristics."[26] But this is not correct, because to be passed on to their offspring, epigenetic effects do not have to alter the parents' traits or characteristics, only their germ cells; or, alternatively, they may signal something about the environment to their developing offspring.

But studying these ideas in humans is not easy. To do so requires investigating aspects of our lives from early development, even of the lives of our parents before we were conceived. This would be an ambitious, long-term—and expensive—venture. And it would take the vision of an English doctor, David Barker, to make it a reality.

Southampton Babies

In Southampton, UK, in the mid-1990s, our colleagues, led by David Barker, established a pioneering study of children, called the Princess Anne Hospital Study after the name of the maternity unit where the children were born.[27] Umbilical cords normally are discarded along with the placenta after birth, but in the hope of being able to conduct later studies when new scientific techniques might become available, the researchers kept portions of the umbilical cords of these babies, and stored them in special freezers at −70°C. When the children were about nine years old, they returned to the hospital for a sophisticated scan that measured their body composition in terms of fat, muscle, and bone. Meanwhile, colleagues in our laboratories had developed the techniques necessary to study gene methylation in DNA extracted from the umbilical cord tissue. The technology at that time only enabled us to study a few genes. We chose to look at a particular gene involved in fat development and metabolic control. We were surprised to find that the degree of methylation of that gene in the umbilical cord was strongly correlated with the fat mass of the children nine years later. The effect was so large that it was hard to believe. Such correlations should not be believed on the basis of one set of measurements alone—they must be replicated. So we repeated them twice more on two separate groups of children from another cohort, the Southampton Women's Survey, which was the successor of the Princess Anne Hospital Study. These children were aged about six, and exactly the same result was found.[28]

Now we could take these measurements of epigenetic states at birth and look back, to examine what aspects of the children's lives before birth might be associated with those states. In the Southampton Women's Survey, mother's diet had been measured before and during pregnancy. We found that the level of methylation in the gene in the baby's umbilical cord at birth was also related to the nature of the mother's diet in the first third of pregnancy. It showed how variation in nutrition in early pregnancy, which was thought to have long-term effects on the health of the child, could alter fetal development via epigenetic gene methylation, and be strongly associated with the degree of obesity in the child six or nine years later.

Since then we have studied epigenetic effects on many other genes and found similar links to aspects of child development, not only in the United Kingdom but in several other populations around the world.[29] A wide range of studies have now shown similar effects of maternal diet, lifestyle (especially smoking), and body composition—such as obesity—on the next generation.[30] Moreover, although initially such epigenetic studies were criticized as focusing too much attention on the diets and lifestyles of women, we now know that the story likely also applies to males. There is a growing literature in animals showing that manipulating the nutrition or weight of male rodents affects the epigenetic state of their sperm and the physiology of their offspring.[31] MicroRNAs are small forms of

RNA that can regulate DNA function through epigenetic processes: testis- or sperm-derived microRNAs injected into mouse embryos caused cardiac defects in the animals later, and these were transmitted to the next generation through the male line.[32] Together these types of experiment make a strong case that environmental influences on the father can produce epigenetic changes in his sperm and that these can lead to significant functional effects in the offspring, potentially over several generations. Subsequent human studies have also shown epigenetic effects of father's body weight, diet, and lifestyle (especially smoking, again) on his sperm.[33]

Such epigenetic processes are now widely believed to explain the multigenerational environmental effects reported for some environmental chemicals. In rats, some chemical insecticides still widely used in parts of the world caused epigenetic changes in sperm that were transmitted to offspring over several generations.[34] In general, where data are available, the epigenetic effects appear to be limited to a few generations and so to be transient, at least in evolutionary terms. But this is what we would anticipate if epigenetic inheritance mechanisms evolved to provide adaptive benefit in many individuals rapidly in response to an environmental change, perhaps giving a survival advantage until more traditional genetic selection processes caught up. They would also be very useful for meeting the demands of environmental changes that are transient, when such genetic selection processes could be inappropriate as well as too slow.[35]

These intergenerational effects challenged accepted scientific dogma but have now been shown under a range of circumstances, including historical events. Women who were subject to famine in Holland during the Second World War have grandchildren with altered hormone levels, metabolism, and epigenetic profiles.[36] But when intergenerational effects are transmitted through the female line, it is more difficult to be sure that the effect is mediated by processes of inheritance—that is, passaging through the egg or sperm to the next generation. The reason is that a woman's ovary forms all the eggs it will ever possess while *she* is still a fetus. So these eggs might receive signals, such as poor nutrition, similar to those received by the woman's mother (the child's grandmother) during her pregnancy even though those eggs will not form the basis for another human being for several decades to come. On the other hand, sperm are made constantly in large numbers by postpubertal men. So if a transgenerational effect over two generations or more is seen to be transmitted by males, this suggests that something intriguing is really going on. We have evidence that this is the case in rats, and there is evidence from some multigenerational cohort studies that similar phenomena might exist in humans. The most quoted study is a historical study of a famine in nineteenth-century Sweden in which cardiovascular mortality in males could be related to whether their paternal grandparents had experienced famine.[37] That study is indirect, though, and for obvious reasons lacks any biological information. But recent studies in humans show the plausibility of male-line-mediated epigenetic inheritance. Offspring of obese fathers have altered epigenetic profiles and their sperm also show epigenetic changes.[38]

We are not saying that all inherited processes have either a genetic or an epigenetic component. The same environment can affect each generation of a family *de novo.* And how parents live influences how their children learn to conduct their lives. Parents who eat a lot are likely to have children who eat a lot. Ciaran Forde, who works with us in Singapore, and his colleagues have shown that children learn concepts of the appropriate portion size for a meal, or a snack, from their parents at an early age, so obesity may simply appear inherited because parent and child share an environment.[39]

Nonetheless non-genomic inheritance over a few generations clearly exists. This has led to big debates. Some neo-Darwinists maintain that the epigenetic observations, while real, are oddities and do not necessitate revision of the classical model of genetic inheritance and evolution. They explain these observations away as being largely irrelevant to the story. But a growing number of influential biologists see these observations as another reason to expand our evolutionary thinking to reflect the variety of potentially adaptive mechanisms in operation—whether over a single generation or several, and involving more than just fixed genetic "blueprints."[40] Moreover, because epigenetic changes can affect the characteristics of many individuals in a population simultaneously when they are all exposed to a similar environmental change, they can provide a powerful transient adaptive mechanism for the population. As epigenetic changes are rapid, they can offset the "adaptive lag," the delay in achieving a fixed genetic adaptive change through selection for genetic variants, which generally requires many generations after an environmental change.

Early Learning

We could ask whether biological information passed transiently across generations could confer adaptive advantage, as defined in evolutionary terms. In other words, could something happening to a parent signal to her / his offspring that there would be a survival or reproductive advantage in altering their biology in some way?[41] We are now in the territory of evolutionary developmental biology, or "evo-devo" as it is called.

Stunting is a form of growth failure that occurs due to severe and chronic undernutrition, often confounded by infection, in pregnancy and infancy. The great UK nutritional scientist John Waterlow pointed out that being stunted is a logical and evolved adaptive response to malnutrition, because reducing growth meant lower nutritional need and thus the child was more likely to be able to survive to reproduce even though there are other biological costs—made worse in the modern world.[42] The child invests fewer precious nutrients in growing their skeleton and muscles, in order to preserve them for the brain, heart, and other vital organs. Sadly, too many children in Africa and Asia still grow up stunted, and the prevalence of stunting is an important public health index of nutrition and child health.

In the 1980s David Barker and colleagues found a cohort of people in Hertfordshire, then in their sixties, for whom birth records were available.[43] Their groundbreaking studies showed that human babies born small were more likely to succumb, as older adults, to chronic diseases such as heart attacks, strokes, high blood pressure, and diabetes. With the Cambridge clinical biochemist Nick Hales, Barker proposed that being born small was most likely to be a result of fetal undernutrition—whether actual undernutrition of the mother during pregnancy or a reduced flow of nutrients and oxygen from mother to fetus due to placental insufficiency. Barker and Hales further proposed that the offspring would develop insulin resistance, which can be a precursor to type 2 or maturity-onset diabetes, as an adaptive response to help reduce nutrient demand and thus promote survival—insulin being the hormone that primarily drives amino acids and glucose into muscle cells. Thus, insulin resistance, like stunting, would aid short-term survival although it might have long-term costs. Barker and Hales joined these three dots together to devise what became known as the "thrifty phenotype" model.[44]

Our own empirical and theoretical research has built on these early observations and led us to further refine Barker and Hales's model. We discussed this research in our previous book, *Mismatch: Why Our World No Longer Fits Our Bodies*. First, we recognized that there are multiple pathways by which exposures in fetal life could have lifelong effects on the offspring.[45] For example, mothers who develop diabetes during pregnancy are also more likely to have children who grow up to have type 2 diabetes: these children are also often born fatter than babies of nondiabetic mothers. Second, we had to explain why insulin resistance as proposed by Barker and Hales was not apparent at birth but appeared only in later childhood. And third, we recognized that the fetal priming of future responses was not a matter of extremes—babies of normal birth weight whose mothers were exposed to malnutrition were also more likely to develop obesity and diabetes later.[46]

Predicting the Future

As we thought more about it, it became clear that the embryo, fetus, and newborn infant are demonstrating an ability to use information obtained from the parents, especially the mother, to "predict" its future environment. It changes the development or function of many organs accordingly.

There are many examples of such predicting or forecasting in a range of species. Pennsylvania meadow voles born in the fall have a thick coat of fur, ready for the winter; those born in the spring have a thin coat of fur. Yet these voles are born after a pregnancy of about twenty-one days in a nest where the temperature is about the same in spring or fall and the mother's body temperature will have been the same during pregnancy. We now know that it is the shortening or lengthening of the day as winter or summer approaches that changes the pattern of production of the hormone melatonin by the mother's pineal gland at the base of her brain. Melatonin is the hormone that rises at night and helps us know it's time to go to sleep. This hormone crosses the placenta and thus signals night length to the developing pups and so whether winter or summer is coming.[47]

Even more striking examples can be found in insects. The desert locust responds to signals about population density secreted by the female in the fluid surrounding the eggs she deposits in the sand. A signal indicating overcrowding will shift development from the green, solitary form, which eats only some plants and does not have to move far to get food, to the darker, gregarious, and more omnivorous form that swarms to find food and can devastate large areas of vegetation or crops.[48]

These developmental responses usually involve trade-offs, because following one developmental pathway excludes others, just as the ball rolling through Waddington's epigenetic landscape cannot change direction. For example, a spadefoot toad tadpole hatching into a desert pond overcrowded with other tadpoles "predicts" that the pond is drying up, as there clearly is not enough water for them all. This prompts its early metamorphosis, leading to an adult toad that can survive to reproduce sooner but is smaller and more likely to be eaten by predators.[49]

From examples like these we came to recognize that environmental signals affect animals, including humans, during development to alter their characteristics for potential adaptive advantage later in terms of survival and reproduction. We called these *predictive adaptive responses*.[50] They contrast with an organism's responses during development to increase immediate survival advantage. For example, the fetal growth retardation that David Barker had focused on is most often an immediate adaptation to cope with inadequate nutrient supply to the fetus, perhaps as a result of the mother being malnourished or her placenta not functioning optimally; the fetus must trade off growth of its body to conserve its energy supply in order to support the growth of vital organs such as the heart and brain. These infants are insulin sensitive so as to get the best nutritional value from human milk and lay down fat to give them buffering stores of energy for the developing brain, because in our evolutionary past weaning was likely to be a nutritionally challenging time. But once the child is weaned, growing up to be smaller and insulin resistant is, as John Waterlow argued, an adaptive response to an anticipated lack of nutritional resources. Of course, if the world is not as bad as predicted, and the offspring face an energy-dense world, then they are at greater risk of developing type 2 diabetes than had they not experienced poor nutrition *in utero*.

We now have more direct support for this idea of predicting, or "forecasting" as the great animal behavioral scientist and evolutionary biologist Patrick Bateson, called it.[51] Working in Jamaica with the clinical scientist Terrence Forrester, we studied infants who had been terribly and acutely undernourished in the first two years of life. Infants who are subject to famine can follow one of two paths. On one path they literally waste away. The condition is called marasmus, and the affected child's fat and muscle break down to provide a source of energy in order to survive. On the other path, the children lose some weight but then their metabolism begins to falter and their tissues fill up with fluid, giving them a characteristic potbelly appearance; their skin tends to break down, and they are susceptible to infections. This condition is called kwashiorkor, and it is often fatal even though these children have not used up all the energy stores within their fragile bodies. In Jamaica many of these children have been rescued and nutritionally rehabilitated, first by John Waterlow and later by Forrester and his colleagues. To their great credit, very few Jamaican children now die of these conditions.

We wondered why a child would follow one pathway rather than another when faced with the same challenge, especially as the kwashiorkor pathway is so much more dangerous. Could it be that the children who developed marasmus were actually different from those who developed kwashiorkor even before they were undernourished?

Forrester and his team tracked down the birth records of 1,336 babies who had later developed either marasmus or kwashiorkor. It turned out that on average the children who developed marasmus rather than kwashiorkor were born with a much lower birth weight. We realized that because these children had been less well-nourished in the womb, they had developed a physiology better able to cope with malnutrition. They had been able to predict the risk of later famine even before they were born, reduce their growth just as Waterlow had proposed, and adapt their metabolic settings accordingly. When we were able to study some of these survivors of famine as adults, we discovered that those who had followed the marasmus pathway had a propensity to become diabetic even as young adults, and remained shorter in stature than the average for the population. When we studied their muscles, we found epigenetic changes best explained by the antenatal experience rather than by their postnatal malnutrition.[52]

It is not just nutrition that can have such long-term effects—stress does, too. Stress in pregnancy in a rat can be mimicked by giving the female a high dose of the stress hormone corticosterone or a potent analogue, and this has long-term effects on the offspring's metabolism. But perhaps the most interesting work comes from studying mothering activity of rat dams with their pups. Suckling dams show a range of behaviors—some lick their pups a lot and some do not, called high- and low-grooming mothers. Scientists in the United Kingdom, United States, and Canada, in particular Michael Meaney and his colleagues, have shown that low-grooming dams have pups that, if female, themselves grow up to be low-grooming dams, and they also have altered hormonal stress responses. High-grooming dams have pups that grow up to be high-grooming dams with lower stress responses.[53] The effect cannot be genetic, because if female pups born to a low-grooming dam are cross-fostered at birth by a high-grooming dam, the pups grow up to be high-grooming dams. Conversely, when pups born to a high-grooming dam are cross-fostered by a low-grooming dam, they grow up to be low-grooming dams. And Meaney and his team went on to show that this is associated with epigenetic changes in the stress pathways of the pups' brains which appear to be mediated by the dam's grooming behavior.[54]

We see something similar taking place in wild animals. Records going back over a hundred years for the trapping of snowshoe hares, and the lynxes which predate them, from the Hudson Bay Company in arctic Canada show that there were large cyclical variations over several years in the number of pelts trappers brought in for sale. There are various explanations for this phenomenon. One is that when there are more hares, the lynx population also increases, after a delay. This reduces the number of hares and so, again after a delay, the number of lynxes declines. Fewer lynxes means that the hare population will increase again, and so on. Studies have now been done on the stress hormone levels in the snowshoe hare. When the lynx population is high, it pays for the hares to be extra vigilant. Stress hormone levels in the pregnant hares will be higher, and this might signal to their unborn pups that it is a dangerous world out there. So the pups grow up with exaggerated vigilance and stress responses, making it harder for the lynxes to catch them. When the lynx population declines, the hares do not have to be so vigilant, so their offspring may be less vigilant too.[55]

Do similar processes affecting behavior occur in humans? Here the studies are more complex because of our long life span and the limited long-term data available. But we do know that children who are physically or mentally abused early in life are more likely to grow up to be abusers themselves. Some limited data suggest that the same molecular epigenetic changes in the brain that Meaney and his team saw in the rat are seen in the brains of suicide victims who are known to have been abused as children. More recently the team, in very detailed studies in Singapore, has been able to show that those mothers with low mood scores (but who are not clinically depressed) during pregnancy have babies with alterations in structures in their brains that in turn affect mood and behavior.[56] This is a very important result, because attention to maternal mental health, unless it is extremely affected, is not considered a public health priority; yet it is clearly of far greater importance than is generally believed.

In this chapter we have discussed ideas about what makes us what we are in terms of what we inherit from our parents and from previous generations. Some ideas have been controversial for very many years, and disagreements about the relative emphasis to place on particular factors within the context of evolutionary understandings continue to this day.[57] At their simplest these disagreements boil down to conceptual understanding of our history. That history has roots in our evolution, and therefore in the lives of our ancient ancestors. It also involves the much more recent past, the early development of each of us and the possible influence of members of our families going back a few generations. What mattered in the past, and what matters to us today, is how well these aspects of our nature allow us to cope with the environment in which we live, and whether we chose that environmental niche, modified it to suit ourselves, or found it imposed on us. Our environmental niche is very much determined by our culture—or, we could say, is a fundamental part of our culture. So now we must turn our attention to that culture, and to the question of whether that too evolves.

4 | CULTURE

IN 1833 HMS *BEAGLE,* with Charles Darwin onboard, arrived in Tierra del Fuego in the far south of the Americas. They had on board three Fuegian Indians who had previously been taken to England to be "civilized" and had lived there for more than a year. They were now being returned home to help set up a mission as part of the great colonial effort of converting "primitive" peoples to Christianity. Darwin wrote about the extraordinary differences in behavior and culture between the naked savages on the beach and the civilized people of Europe. He continued to reflect on these differences over the next fifty years.[1] This played a considerable role in the development of his thinking about the evolution of expressions and emotions in animals and humans, which he wrote about in his book *The Expression of the Emotions in Man and Animals* (1872).

The *Beagle's* voyage was only one of many journeys of discovery in the nineteenth century. Darwin was inspired by the journey to South America of the great German polymath Humboldt, who had written extensively on his voyage of discovery—books Darwin took with him on the *Beagle.* At the beginning of the nineteenth century, European science and society were intrigued by the diversity of the natural world, as reflected in the urge to collect specimens (which financed Wallace's expeditions), but also by the diversity of cultures. We can trace the beginnings of the sciences of sociology, anthropology, and ethnography to this period. A good example is Richard

Burton, the Oxford-educated Arabist, linguist, and explorer who later co-founded the Anthropological Society. Burton started his career in India, where, among other things, he tried to converse with monkeys. But he became famous for his journeys in Arabia and for seeking the origin of the Nile. In 1853 he was one of the first non-Muslims to participate in the Haj pilgrimage to Mecca (which he did in disguise). His extensive writings detailed the many different customs of the various ethnic and tribal groups with whom he came in contact, including very detailed descriptions of their sexual practices.[2] His translation of the *Kama Sutra* and other erotic writings gave him both fame and infamy.

Soon after, in 1877, the American anthropologist Lewis Henry Morgan was completing one of his most influential books, *Ancient Society, or Researches in the Lines of Human Progress from Savagery through Barbarism to Civilization.* Morgan had championed the cause of the Iroquois nation, living in the Northeast of the United States, who had been badly treated by colonists for many years. When he realized that the Seneca and Ojibwa tribes ascribed kinship to individuals within their societies in very similar ways, and very differently from Europeans, he developed a theory that the origins of these American peoples were more closely linked to Asia than to Europe. His studies led him to consider how cultural beliefs and practices are passed down from generation to generation, and also how they change over time—reflections that led him to develop the concept of "cultural evolution." At the end of the nineteenth century another father of sociology, Émile Durkheim, was to consider how societies could maintain their structure and function as they went through "modernization" and the weakening of religion and social ties, which he considered the unifying bases of societies. So it was clear that societal structures depended not only on kinship and genetic relationships but also on cultural bonds that hold groups together.

Through these emerging disciplines of evolutionary biology, sociology, and anthropology, we came to understand that culture is central to what makes us what we are. But what is culture? The word can be taken to mean many different things. In the context of this book, *culture* includes socially defined and constructed behaviors and organizations and the products of our collective ingenuity and creativity in arts or sciences, and in particular in the material cultures of tools, machines, and technologies.

Evolving Culture

We tend to think of culture as a uniquely human characteristic, although there are manifestations of what we might call culture in the group behavior of other primates and even other large mammals such as cetaceans. Gorillas communicate, they use primitive tools, and they have complex behaviors. In Rwanda, the new silverback male we met quickly learned from the females of his new harem, who had been accustomed to meeting human tourists, that we did not pose a threat. An aspect of his behavior had changed rapidly.

We see human culture as being qualitatively different, however. It is not constant, it evolves. We understand human culture to include the capacity to teach and to learn, to communicate complex ideas (including those about death and the supernatural, leading to many customs and to religion itself), the ability to think in abstract terms, the capability to make things such as tools, and the facility to refine and improve technologies over time. The interplay between the evolution of the hand and opposable thumb, of language and of a big brain, is still debated by evolutionary biologists. Toolmaking started at least 2.5 million years ago, and possibly as far back as 3.3 million years ago.[3] But it evolved very slowly until perhaps 50,000 years ago, about the time when there was also an explosion in artistic and ornamental expression. How and why did an aesthetic sense evolve? Did we evolve an appreciation of beauty for its own sake?[4] Or because it conferred a specific adaptive advantage?[5] Did an appreciation of nature result from a sense of beauty? Or is it simply a way of assessing the suitability and safety of environmental conditions? This debate has parallels with Wallace and Darwin's debate about sexual selection.

The origin of language is also hotly debated. Estimates of how long ago it emerged range from hundreds of thousands of years ago to just the last 50,000 years.[6] But the fact that human language exists, in contrast to the forms of communication used by even our closest biological relatives, such as chimpanzees, is taken as evidence for the concept of gene–culture coevolution, which we discuss later in this chapter. The physical differences between chimpanzees and humans make this clear. Compared to the chimpanzee, in order to speak as we do, our larynx must be positioned differently in the neck and have a very different structure and our tongue must be far more mobile. In our brains, the areas associated with speaking and understanding the speech of others—the Wernicke and Broca areas—are larger than those of our cousins. Clearly, then, there are genetic differences between the chimpanzee and humans that have evolved in the service of speech. But there is more; we learn to speak, and in a particular language, from our parents and others. And this involves mimicry, practice, and a degree of trial and error—all the learning processes of the brain and the body that are fundamental to growing up in a society.[7]

Culture evolves by complex and dynamic mechanisms. For example, it can be inherited in the form of language, religion, rules, or political affiliation. But the line of inheritance is not strict. Orthodox Jews may be the children of Orthodox Jews but this does not ensure that their children will be Orthodox in belief or practice. They may adopt another religion or cease religious practice altogether. Cultural practice and belief is not influenced solely by our ancestors and parents, it emerges from the milieu and society in which we live. And it can change quickly. Jeremy Button, one of the three Fuegians who were returned to Tierra del Fuego on HMS *Beagle* after time in Britain, had adopted British dress and the customs of Europe, including Christianity.[8] But after returning home he reverted to the culture of Tierra del Fuego and may have later participated in the murder of missionaries. Where cultural traits resonate within a social group, they will persist or further evolve, but where they do not (as Google Glass did not) or where they are displaced by something else (as the Walkman was), they can quickly become extinct. And cultural evolution has few constraints—the major one being whether sufficient numbers of people in a society see the value in an innovation. Advertising and democratic politics are in no small part based on persuading enough of us in our society to sustain or adopt a culturally evolved trait.

Biological evolution is constrained to an extent by the genetic substrate of the past, but this is not necessarily true of cultural evolution. Morse code was replaced by speech transmission by radio, telephone wires by fiber optic cables and microwave transmission. Going to the cinema is being replaced by streaming services like Netflix. Biological evolution proceeds by small, incremental changes. But because each change affects other aspects of biology, and because the key to such evolution is the survival of the organism to reproduce, biological evolution is constrained in time and by the range of possibilities for a particular line of inheritance. For example, evolving a bigger brain to incorporate larger speech and language areas has to be traded off against the limits on energy intake needed to support a bigger brain. There are no such inherent constraints on the possibilities for cultural evolution. The constraints that do exist are those of culture itself, and these can change rapidly. Consider how rapidly, in most liberal countries, attitudes have shifted regarding gay marriage—from its being a religious (cultural) taboo to its being an accepted human right.

But just as biological inheritance is subject to evolutionary pressures, so too is cultural inheritance. Languages clearly evolve over time—French, Spanish, Catalan, Romansh, and Italian all evolved from Latin. Religions evolve over time too, as do beliefs and social constructs. Concepts of privacy are a good example: England has moved from the prudish values of Victorian society, where even the legs of tables were covered, to explicit sexting of the most intimate parts of bodies on social media. We can see cultural evolution in all domains of human endeavor: art, music, fashion, the status of women, our attitudes to human rights, our attitudes to our environment. Science itself also undergoes cultural evolution as there emerge new technologies that in turn create new possibilities for knowledge generation: areas of science become more or less fashionable, something like trends in popular music.

But there are features of cultural evolution that differ from those of biological evolution. First, cultural change can happen quickly for many reasons and without constraint. Look how quickly fashions change or new words like "post-truth" can enter our lexicon, and how fast technologies are displaced. Who now would use a slide rule? Many children reject the religion of their parents and the rituals associated with it, and in England church attendance has fallen dramatically within a generation.

Second, whereas biological evolution can only pass directly from parent to offspring, cultural change can come from peers or unrelated individuals. Biological evolution is subject to clear definitions of what leads to success—namely, survival to reproduce within the lineage. Cultural evolution is much more capricious—fur coats, once fashionable, now are unacceptable. The music industry thrives and music evolves because young people pick up their likes by reference to the views of their peers, not their parents. Why did the small sect that Jesus led in Judea have ultimate success while many other messianic sects that existed in that region at about the same time did not? Why were ultra-orthodox Jews able to sustain their identity in a totally alien environment (northern Europe) over several hundred years? Clearly there are both internal reasons (strength of group identity and value) and external reasons (persecution, confinement to ghettos, variable political support) that affected their survival.

Third, as culture evolves, it can accumulate complexity. Part of the reason for this is that cultural transmission between individuals or from one generation to the next inevitably involves some modification. The continuous ingenuity needed to generate new aspects of culture as older ones are replaced has been likened to running on a treadmill to stay in the same place. This process speeds up as population size increases and there are more inputs into innovation. The treadmill model has been contested.[9] But we find echoes of it in the increasing speed of innovations and cultural evolution associated with IT-based technologies, which we will discuss in Chapters 7 and 8.

Cultural evolution can have runaway characteristics. We met the concept of runaway selection in Chapter 2 in relation to sexual selection, which favored the evolution of enormous antlers in the Irish elk. Innovation accumulates on innovation. Cumulative social innovations like the development of schools and formal education, the development of the nation state with its implied contract between government and citizen, the expansion of collective knowledge through spoken stories, then books and libraries, and now the internet, have created conditions where the capacity to innovate, to project innovations, and for these to have broad impact, is rising exponentially.

The issue of climate change demonstrates this acceleration. In the preindustrial era, human impact on global climate was insignificant in comparison to natural climatic variations. Deforestation and methane production from rice cultivation in paddy fields would have led to some marginal changes in greenhouse gas levels in the atmosphere. But in the nineteenth century, with industrialization and the rapid development of fossil-fuel-based energy production and consumption, carbon dioxide emissions started to rise. Further, with innovations that improved public health and nutrition, the rate of growth in the global population accelerated exponentially. Industrial manufacturing and developments in transport and the motorized vehicle led to massive increases in CO_2 production and nitrogen emissions. Food production to meet a growing population led to accelerated deforestation and further increases in carbon dioxide production. Pastoral farming for beef and milk production added substantial amounts of methane and carbon dioxide to the atmosphere. Fertilizers added nitrous oxide to the mix. All of these human activities place the world in an increasingly perilous and warming state: a consequence of cultural evolution.

A distinctive feature of our more recent cultural evolution, which is well demonstrated by the issue of climate change, is that the populations most affected by any harmful effects are not necessarily those which originated the environmental change. The increase in sea levels from global warming will result from the activities of large industrialized countries but will have the most devastating effects on poor coastal communities and small island states with much less industrialization. The potentially damaging effects of the digital media platforms, which we will discuss in Chapter 8, are similarly remote from the innovators. Such challenges illustrate a fundamental point about human cultural evolution and niche modification. Many species construct a niche to promote survival, by gaining advantage over competitors and minimizing the effects of variations in the environment. By doing so they achieve an evolutionary equilibrium with the local environment. In our case, the innovations are made for different reasons, for pleasure or profit for example, and we do not aim to achieve an equilibrium with our environment. Rather, we continually modify it—or perhaps we should say that some of us do, even if many others feel the consequences, intended or not, good or bad. Human-mediated niche modification thus has the potential to have much broader consequences than niche construction, where the effects are limited to the local ecology.

Tools and Technology

Animals learn from one another, sharing information: gorillas can use sticks as simple tools, and chimpanzees fish for insects in trees using sticks. Macaque monkeys in Japan have learned how to wash potatoes by watching and copying one another.[10] But despite these examples, there is a massive quantitative and qualitative difference between these characteristics in animals and in humans. Our cognition, manual dexterity, learning, and communication give us the unique abilities to innovate continually, building on previous information and endlessly modifying our niche. These abilities make cultural evolution critical to understanding the human condition.

Humans are persistent and cumulative innovators of technologies. Our success at this depends partly on the social environment in which we live, the cognitive and language abilities we have, and particularly our ability to observe, learn, and imitate each other. Our technological ingenuity in many ways defines our species. Technological evolution appeared to be slow in our evolutionary past until perhaps 50,000 years ago, but then it started to accelerate. Some argue that this is evidence that language did not evolve until then—perhaps due to changes in the brain from genetic mutations.[11] Others argue for a much earlier origin of language. Whatever the underlying reason, gradually the pace of technological change started to increase; pottery, bronze, the wheel, advanced weaponry such as armor and catapults were developed. Then came the invention of printing, then the Industrial Revolution, then the manufacturing revolution, then the development of motorized transport on land and sea, flight, the telephone, electricity, nuclear power, pharmaceuticals, nuclear weapons, computers, fast food, genetic technologies, and the internet. In every case the cultural evolutionary history is clear—each innovation builds on past innovations, complexity increases, and the pace of invention accelerates.

To Be Human

We humans have a remarkably evolved substrate for our behaviors in the form of our greatly expanded neocortex—the outer layer of the brain, which is primarily responsible for our cognitive and related skills. Yet even back in the early debates between Darwin and Wallace and Owens and Huxley, the evolution of human behavior was one of the most contested areas of evolutionary biology. Part of the reason is that human behaviors are qualitatively so different from those of any other animal that experimentation and extrapolation from animal research is difficult. And because many of the tools of evolutionary science simply cannot be used in this field, there is no way we can be sure what the behavior or thought processes of our ancestors were like 30,000 years ago.

Several schools of thought, based partly on different conceptual approaches, have emerged regarding the evolution of human behavior and brain function.[12] At the extremes, two views exist. One is that our brain evolved as a series of domain-specific modules. The second view is that our brain is a general "tool" that is able to respond flexibly to many situations.

The most prominent advocates of the modular view are evolutionary psychologists Leda Cosmides and John Tooby at the University of California, Santa Barbara.[13] They hypothesize that there was such strong selection pressure in our ancestral past that each capacity of the mind has evolved as an independent module. They argue that there are perhaps many thousands of modules, each for a different behavior, such as a module to detect freeloaders, a module to learn language, and so forth. A key concept in their thinking was that modern humans underwent selection within the "environment of evolutionary adaptedness"—the environment that is believed to have existed through the majority of human existence, at least until the end of the Neolithic period, during which selective pressures acted on human biology to lead to the current portfolio of human behaviors. The modular model implies that behaviors were selected when we were hunter-gatherers living in small groups and that they must largely be genetically determined.

Stephen Pinker and Robert Wright are two of the scholars who have popularized these concepts.[14] They suggest that the environment of evolutionary adaptedness was the Paleolithic world, and that at this time the fundamentals of human culture were laid down. They would include the origins not only of our family structure and tendency to settlement, but also of social norms, sexual behavior, altruism and cooperation, social hierarchies and aggression toward rival ethnic groups, and gender differences.[15] These complex behavioral "modules," they argue, would not have evolved had they not conferred a fitness advantage. And because they have evolved, they must be the result of natural selection of a range of heritable, genetically determined traits.

Despite the widespread popularity of these concepts, there are many problems with them, well summarized by the Exeter philosopher of science John Dupré. Why, asks Dupré, if we take a strictly adaptationist view of our biology, would we pick the Paleolithic as a period in our ancestry to which we were best adapted?[16] After all, our predecessors' species spent far more time (billions of years) as unicellular and simple multicellular organisms. Why weren't we best adapted to Pleistocene or Cretaceous conditions? At the opposite end of the story, why are we genetically 98 to 99 percent identical to our nearest cousins, the chimpanzees, with which we share the last common ancestor before our lineages diverged several million years ago? Our behavior and many other aspects of our biology, from the scale of our language acquisition to the length of our large intestine, are very different from chimpanzees. This makes it hard to envisage the difference in terms of selection of complex genetically determined modules of behavior. Further, as the Cambridge anthropologist Robert Foley points out, there was no single environment of evolutionary adaptedness.[17] By 50,000 years ago, humans lived across Africa and much of Europe and Asia and had reached Australia. Different groups of humans would have met very different environmental challenges that would have required very different behaviors for survival.

Dupré further argues that we are more innovative than the idea of Stone Age brains would suggest. Where did our abilities to mount the Industrial Revolution, let alone to travel to the moon or invent the internet or gene editing technology, come from? If we are fundamentally adapted to life before such things, where did the ingenuity to develop them arise from, and how was that selected in the Paleolithic as adaptive? Those Stone Age brains must have been very prescient indeed! Or to put it another way, given our success as a species since the Paleolithic, this theory seems to contradict all of the conceptual arguments on which this school of evolutionary psychology is based.

The protagonists of the "brain as a general tool" model, the second view, suggest that most behaviors are learned, but that they can be learned only because of the evolved neural substrate of the brain. They argue that humans evolved with a neural infrastructure highly capable of learning from experience and from others, and with capacities to cope with novel situations and live within complex social organizations.

The issue of the evolution of language highlights the debate. The linguist Noam Chomsky argued along the modular lines and suggested that there must be a "grammar" module which underlies all languages.[18] Yet linguistic studies have failed to demonstrate a common grammar substrate—there are too many exceptions, even though all humans have evolved with an ability to use language in thought and communication.

Because early human brains and bodies were similar to ours today, some might leap to the assumption that their emotional responses to situations were similar to ours under similar circumstances—if such circumstances ever exist today—and that their ambitions, hopes, and fears were similar to ours. This is evident in the writings of authors such as René Dubos, who considers the challenges we face as humans living today with bodies that evolved in the Paleolithic.[19] Accounts like this raise issues about interpretation. By describing processes of cultural evolution, when changes in the habitat of ancestral peoples were accompanied by changes in behavior and technological advances, it is possible to imagine how their lives changed, but were there accompanying changes in how their brains functioned? It seems likely that the answer is yes. For example, we know that brain function is affected by how we use our brains. Taxi drivers in London, who must undergo a long period of training to memorize the city streets, have larger hippocampi—areas of the brain associated with memory—than do bus drivers, who may cover as many miles but follow set routes.[20] There are data suggesting that brain structures are affected by the complexity of the networks in which we live and operate.[21] Some data suggest that children's brain structure and attention spans may be affected by their screen-based technological environment.[22] And as we discussed in Chapter 1, genetic studies suggest ongoing selection in genes related to brain function over the last 10,000 years.

While we broadly disagree with the modular model as the basis of understanding human behavior, it does emphasize an important point. The human brain evolved in social and macro-environmental conditions that are very different from those in which we now commonly live. If these modules were based on psychological adaptations appropriate to the conditions in which they evolved, then there would now be a mismatch between those modules and the modern constructed world. There would therefore be situations where the adaptations that underlie human behavior have lost their adaptive advantage, and might now manifest as maladaptive pathologies. Fundamental to this school of thought is understanding the circumstances that led to a particular module of behavior being selected. For example, one might envision a module for fear of dangerous animals, such as snakes. The evolutionary psychiatrist Randy Nesse, who is one of the founders of the field of evolutionary medicine, would suggest that echoes of such thought processes can give rise to pathological phobias in our modern world even if we never encounter any snakes.[23] Then again, some modules may function in ways that reflect aspects of life that have changed very little: the emotion of jealousy might result from a module that had evolved through conferring an adaptive advantage, because a jealous individual might be more likely to have a selective fitness advantage over someone who took a passive view of being the victim of infidelity, and the emotion may persist today even though the adaptive advantage does not.

Among Friends

While our understanding of the evolution of our brain and behavior largely assumes that it is based on individual adaptive advantage, we must also consider how our societies evolved. Humans have always lived in groups. Not all primates do. Orangutans come together to mate but otherwise live solitary lives. We can speculate that group living was essential for the success of hunter-gatherer clans and was of considerable value in supporting our long, dependent period of infancy and childhood. The presence of an extended family, and the maternal grandmother in particular, aided maternal fitness because the mother was more likely to have more children who would survive to adulthood. One of the favored hypotheses for the evolution of menopause in humans is that there was a major fitness advantage in having a post-reproductive grandmother to help her daughter and grandchildren thrive.[24] Some recent data suggest that such an effect still exists. Group living also provided the basis for collective and extended knowledge, learning, and innovation.

As agriculture, settlements, and the differentiation of tasks within society developed further, group identity extended beyond close kinship to more distal and unrelated individuals. Groups require cohesion, and cohesion requires rules—religion, laws, custom, and material culture all help to define a group. They clearly offer identity and establish what is acceptable within a group and what is not. Breaking the rules of the group or not pulling one's weight by free-loading would lead to punishment or exclusion from the group. In the past, exclusion from the group would have often meant death.

One major area of ongoing debate regards the origin of altruistic behavior—our willingness to assist another member of our group without expectation of immediate reward. Altruism is very much a feature of group identity; we are more likely to be altruistic within our own group. Numerous academic books have been written about it.[25] Two main lines of argument have emerged. The first argues that altruism is based on kin selection and originated in Paleolithic groups of related individuals. The evolutionary geneticist J. B. S. Haldane is reported to have said that he would gladly give up his life for two brothers or eight cousins (although others have attributed this statement to the evolutionary biologist William Hamilton).[26] Regardless, genetically deterministic evolutionary theory argues that the reason for altruism in support of the survival of close relatives is that they share greater commonality of genes. The alternative theory argues altruism is based on reciprocity—a member of a group will act for the good of other members—whether related or not—because that creates a group debt to be repaid if the individual needs help in the future. Both arguments, which are not as distinct as they first appear, gave rise to the field of evolutionary game theory, which modelers have developed in order to explore how one might gain advantage in a group and how one might detect a freeloader.

If such behaviors are genetically determined, then selection of a particular behavior may be favored within a group and promote reproductive success. This could occur at an individual level and become more frequent within a group. Groups would also expel those who did not conform, and thus cheaters or freeloaders would not flourish. Religions excommunicate dissenters; legal systems put cheaters in jail or fine them. As a group becomes more cohesive, with common behaviors, this might favor the whole group over a competing group that was less cohesive, and soon such cohesive groups in a region occupied by that species would predominate.

The first two arguments—kin selection and reciprocity—were seen as sufficient to explain altruistic behavior by evolutionary biologists who rejected the group selection argument. Today a more nuanced view of multilevel selection is more widely, although not universally, accepted—namely, that group selection, individual selection, and kin selection are not mutually exclusive and that under some conditions group selection can occur whether at the biological or cultural levels (see also Chapter 3).[27]

Genes and Culture

The relationship between cultural evolution and biological evolution is clearly complex. As discussed, our biology allows us to have cultural developments and characteristics—behavioral, social, and material—that, compared to any other species, are very different in degree, type of effect, and plasticity. On the other hand, our cultural developments affect our biology. As our hominin ancestors learned to control fire, they became skilled at cooking. Cooking had the advantage of making raw foods, especially meat, less tough. As we used cooking more, the need for powerful chewing muscles declined and the genes that controlled jaw growth became less selected. Our jaws are much less protruding than those of our ape cousins, even those that are vegetarian. Greater intake of protein also drove increases in stature throughout most of our history although, for reasons we will discuss later, human height fell after the development of agriculture and permanent settlements, and only recovered beginning in the nineteenth century.

Perhaps the best-known example of this gene–culture coevolution is our ability to digest cow's milk. Human survival depends on our ability as infants to digest milk, because until recently, in evolutionary terms, the only source of infant nutrition was mother's milk. Digesting milk requires lactase, an enzyme in our intestine, to digest the sugar lactose in milk. Without lactase, which some people lack after infancy just as all our early ancestors did, drinking milk produces abdominal discomfort and severe diarrhea. Apart from the inconvenience, this is a major disadvantage in a situation where animal milk provides a good source of protein and energy.

In premodern times there was no need for the enzyme lactase after weaning because there was no nutritional source of lactose, and so we evolved with the gene for lactase being naturally turned off at about three years of age—the age when our ancestors likely weaned their infants. Then humans began to domesticate cows about 10,000 to 12,000 years ago in the Middle East, and cow's milk became a valuable nutritional source. It appears that early pastoralists may have found ways to remove lactose from cow's milk. In the 1970s the archaeologist Peter Bogucki was examining pottery fragments from a Neolithic site in Poland dating to 8,000 years BP. The pottery appeared to have had perforations similar to contemporary methods for filtering whey in cheese making. Later, chemists found residues of milk fat in the pottery fragments.[28] And earlier agriculturalists in the Middle East, about 11,000 years BP, had invented a method for removing lactose from milk by curdling the protein and then sieving it.

Humans in those early dairy-farming communities, who likely ate cheese before they drank milk, chanced to develop a mutation that left the lactase enzyme turned on after infancy. Individuals with this mutation had better nutrition and thus were more likely to survive and reproduce. This mutation spread rapidly through descendants of that population—almost all modern European and Middle Eastern peoples have the ability to digest milk sugars. Those who do not are lactose intolerant.[29]

Independently, cows were domesticated in East Africa several thousand years ago and a different mutation in the same gene, with the same effect, developed. Therefore, many East Africans are lactose tolerant. There has been further evidence of positive selection for lactase persistence in a Chilean population that has recently adopted agro-pastoralism.[30] In contrast, dairy farming was not a cultural innovation in East Asia and so the mutation for lactase persistence did not take hold there, and lactose intolerance is common in East Asia today. This is a very specific example of how our culture has affected our biology and in turn how our biology induced cultural changes, through *gene–culture co-evolution*. This concept was first identified and named by the two important evolutionary thought leaders Robert Boyd and Peter Richerson.[31]

Today, with our ability to sequence the entire human genome in great detail, we are finding more associations between our culture and our genome, including an increasing number of genes associated with digestion that appear to have been influenced by our diets. For example, although we all have variable numbers of copies of the gene for the enzyme amylase that digests starch, those who live in cultures with high-starch diets tend to have more copies of that gene, which makes it more effective.[32]

Humans' ability to live in groups was probably enhanced by genes that influenced human behavior to be fair, altruistic, and eusocial. In turn those abilities enhanced the survival of those in groups of people with favorable mutations in these genes. Whether by individual selection or by group selection or both—as in multilevel selection—and supported by the cultural evolution of mechanisms to sustain group coherence, the social nature of humans was reinforced.

Reflection on these interactions emphasizes the illogicality of trying to separate "nature" and "nurture"—it's a false distinction that has nevertheless been frequently adopted, both in science and more widely, and that still continues to be promoted.[33] Our genes, our development, and our social environment all work together to make us what we are.

Biology and Behavior

Thinking of human characteristics in purely genetic terms, selected over millennia, can lead to some extreme views—such as the view that behavioral attributes have a largely genetic basis and so can be assigned stereotypically to different ethnic groups. This can fuel racist and eugenic arguments, which unfortunately are still part of the platform for some populist movements today and, although widely deplored, persist. Such views are not new. The concept of "civilized" societies having succeeded over "primitive" ones through a process of natural selection, where those perceived to be hereditarily "higher" or fitter triumphed over those less fit, was widely held in Darwin's time and later.[34] Such thinking was opposed by a group of anthropologists led by Franz Boas, whose student Ruth Benedict and her student Margaret Mead explored ways of seeing culture quite differently in their studies of a range of peoples.[35] They viewed culture as much more nuanced, not grounded simply in genetic history, and potentially varying infinitely between one society and another. They did not dismiss the importance of the interaction between human culture and environment, which, they maintained, allowed a diverse range of cultures to develop and coexist in the world: instead, what they objected to was the hierarchical concept of cultures as evolving from "lower" to "higher" forms.

These issues came to the fore when the distinguished evolutionary biologist and thinker E. O. Wilson published his book *Sociobiology* in 1975.[36] Wilson applied neo-Darwinian ideas of genetically fitter populations triumphing over those less fit to the study of behavior in social animals ranging from ants to humans. There were violent campus debates over whether human behavior could be rooted in such biology. Wilson argued that an understanding of human behavior required more than sociology—it needed "consilience," by which he meant an ability to move between disciplines in an integrative fashion. There is much validity in his consilience argument, and it is true that by adopting an extreme position some members of the social sciences community isolated themselves and pushed the discipline further from biology by denying the latter's importance. However, Wilson's intervention came at a time when the neo-Darwinist position of gene-centric deterministic biology was at its height. Unlike some other, more speculative neo-Darwinists, Wilson was a highly accomplished experimental researcher. His work on the behavior of social species, especially insects, is renowned. The problem perhaps lay in his somewhat uncritical application to humans of his ideas about societies of ants, and his fundamental belief that the basis of human nature lies in our ancestral environment. To quote from his book *Consilience,* "the natural environment is the theatre in which the human species evolved and to which its physiology and behavior are finally adapted. Neither human biology nor the social sciences can make full sense until their worldviews take account of that unyielding framework."[37] Wilson's view suggests that genes selected by evolution are the primary determinants of behavior.[38] This biologically deterministic position does not seem to offer hope of a closer collaboration with the more anti-reductionist or anti-adaptationist members of the social sciences community. It does not foster the type of interdisciplinary collaboration essential for understanding what makes us what we are—a point we will return to in Chapter 9.

Behavior depends on a biologically constructed substrate, our brain.[39] This develops and functions as a result of an interaction between our genes, our development, and our environment, and perhaps the most important part of that environment is the group or groups we feel part of and with which we wish to be affiliated. It is essentially impossible to deconstruct that interaction, but by understanding this intertwined perspective we can make sense of recent observations. For example, the distinguished developmental psychologist Michael Meaney and his colleagues have shown in very sophisticated studies in Singapore that children's brain development—and in particular the size and connectivity of the parts of the brain associated with emotional self-control, and so the children's behavior—is affected by maternal mood during pregnancy.[40] The effect is mediated via genes affecting brain growth and function. Consilience is indeed needed to study human behavior.

As many societies face the challenge of maintaining social cohesion in the face of change, understanding how human behavior develops and is affected by developmental and environmental factors becomes increasingly important. The Nobel laureate James Heckman has pointed out that high school graduation depends less on so-called intelligence and much more on emotional and social skills developed in the first few years of life. And his studies go on to show that poorly developed emotional and social skills have longer-term consequences for employment, crime, and relationships.[41]

Just So . . .

In discussing cultural and behavioral evolution, there is a danger of falling into the trap of "just-so stories"—a critical phrase taken from the children's book of the same name published by Rudyard Kipling in 1902, in which he comes up with fanciful explanations for animals' having the characteristics that they do have, without any supporting evidence. How did the leopard get its spots, for example? It is easy to come up with stories, but much harder to prove their veracity. For example, if we look at the literature purporting to explain why humans are bipedal, there are many suggestions, but none is compelling or based on strong evidence. They include the need to keep the hands free to use weapons and tools; to cope with heat in the east African savannah by exposing less of the body to the sun; or to make moving as hunters more efficient. And there are many more speculative hypotheses.

Furthermore, the evolutionary biologists Stephen Jay Gould and Richard Lewontin alerted us to the dangers of believing that every attribute of a species must be adaptive (that is, evolved through natural selection for some evolutionary fitness advantage).[42] Some attributes, they argued, might be just like the spandrels of the Basilica di San Marco in Venice. Spandrels are the architectural result of building an arch, one of the strongest methods of supporting the structure of a building above, into a wall. Because the arch is curved, there will be an area above it on either side which is roughly triangular and has to be filled in to complete the wall. In the Basilica, this space is filled with beautiful decorative stone carving. The carving has aesthetic qualities but is not needed to support the wall. Plain stone would serve just as well.

Another danger, highlighted by Gould and Elisabeth Vrba, lies in not understanding that some features of a species may be "exaptations."[43] These are attributes that evolved as adaptive for one function in a previous context, but are now used for a different function in the current context. The best-known example of an exaptation is feathers, which evolved in dinosaurs for warmth but then were co-opted to assist flight in their descendants, birds. Another example concerns the small bones of the middle ear in mammals, including humans. Over the course of mammalian evolution, bones that made up part of the jaw were repositioned and repurposed to become the ossicles in the middle ear, which are essential for transducing sound from a pressure wave entering the ear into a mechanical movement that stimulates the nerve cells of the inner ear. A fundamental aspect of evolution is that it is a gradual process of change in existing characteristics. So the feathers of birds and the middle ear ossicles of mammals did not arise out of the blue—they were based on reusing and modifying characteristics of foregoing species, rather than total deletion and replacement by novel ones. But how fast is evolution in reality?

Fast or Slow?

The classic understanding of biological—genetic—evolution is that it is slow. It takes many generations for a mutation selected in an individual organism to spread across the population. Because genetic evolution has no direction, the occurrence of the mutation is by chance, and so it may take very many generations to occur, if ever. There are other complexities too: many genes are involved in determining any one trait; many genes have multiple effects; and selection involves trade-offs between traits.

But as we mentioned in Chapter 1, the idea of slow biological evolution is being revised in the light of new evidence.[44] Rapid biological evolution can occur only when there is very strong selection pressure, when the environment is such that only those with certain heritable traits determined by a few genes can survive or thrive. The most generally cited example is that of the finches in the Galápagos which Darwin collected. Peter and Rosemary Grant recognized, 150 years later, that, depending on rainfall, finches with certain beak sizes were more likely to survive. Beak size determined the kinds of seed they could eat, and the numbers of different kinds of seeds depended on whether it had been a wet or dry year. They saw this as an example of strong selection for beak size and very rapid evolution. The more biologists look, the more examples they find of rapid or contemporary ongoing evolution.

Initially it was thought that the genetic changes underlying biological evolution occurred only by chance, independent of the environment, and that even if a change conferred some advantage, it would take many generations to become common in a population, creating a potential adaptive lag—that is, the temporal gap between when the environmental change occurred and when biological evolution had reached a point when a new equilibrium between the environment and the phenotypes of individuals in the population was established. It is now becoming apparent, however, that the epigenetic changes we discussed earlier can affect the rate of mutation in some parts of the genome.[45] Because such epigenetic changes can be induced by a change in the environment, and can affect many individuals simultaneously, this opens up the possibility of more rapid evolutionary change in a population in response to such an environmental change. Epigenetic changes can be produced by aspects of the environment, such as diet or stress, that we would usually view as culturally mediated. So the new science of epigenetics is further blurring the distinction between biological and cultural evolution.

The ability to sequence the whole human genome has enabled geneticists to demonstrate subtle but ongoing genetic change that suggests some ongoing selection at the genetic level within our species, but major differences in traits are difficult to ascribe to these changes. In contrast, extremely rapid cultural evolution is a feature of modern societies. We not only create niches but also continually modify them, through ingenuity, to meet the demands of our incessant drive for cultural evolution. Our new environments then create new opportunities for gene–culture co-evolution. But the changes we are creating through innovation are occurring at such a great pace that it is becoming much more likely that there will be mismatches between our biological evolution and the consequences of our cultural innovation. This will be the focus of Chapters 8 and 9.

The challenges we now face are simply the consequences of the evolutionary path we have followed. We are the only species that continually tries, intentionally or not, to change nature. But this leads to an important further question. Does our changing of nature through our ingenuity cause changes in our biology? If so, what are the implications for our future? To answer this we will have to look more closely at how we have attempted to change nature, in order to see whether nature in return has changed us. We should start with one of our fundamental cultural strategies: settling in one place to live in increasingly larger groups.

5 | SETTLEMENTS

Anthropologists believe that for most of human history, until very recently in evolutionary terms, humans lived in small family-based groups. Living in groups allowed collective hunting in the same way that a pride of lions, a coalition of cheetahs, or a pack of African wild dogs hunts together. And it allowed tasks to be shared—hunting, childcare, defense, the gathering of seeds, fruits, and tubers, and so on.

A range of studies have shown that humans prefer to operate within social groups of a certain size. The best-known studies are those of the distinguished evolutionary psychologist and primatologist Robin Dunbar, who correlated social group size with brain volume in primates and extrapolated this to humans, concluding that a group size of about 100 to 150 is optimal for us.[1] He then examined a range of anthropological, archaeological, and other sources, from census data to records of contemporary hunter-gatherer societies, and so on, arguing that many aspects of human behavior, such as optimal group size, would have evolved during the Pleistocene period. His research suggests that humans have an affinity for a group size of about 150, as this seemed to be the preferred size for units of the Roman Army, Neolithic farming villages, Hutterite settlements, optimal industrial company unit, or academic specialty groups. It is the average size of an individual's personal contact list in their smart phone contacts. Dunbar informally described his number of 150 as the "number of people you would not feel embarrassed about joining uninvited for a drink if you happened to bump into them in a bar." Of course, as Dunbar points out, the number is not the same for all types of social interaction, with 150 being around the maximum for informal interactions with people we know, although perhaps not very well. For more intimate conversations, the preferred group size is much smaller. While 150 people is a comfortable social group for many of us today, early hunter-gather groups were likely much smaller collectives.

Living a Mile High

One of the best-researched hunter-gatherer populations inhabited the Colorado Plateau, an area about 4,000 to 11,000 feet above sea level in the southwestern United States. This lofty plateau was occupied from at least 10,000 years ago. As the climate warmed after the last ice age, humans inhabiting this region hunted giant sloths, shrub and musk oxen, mountain goats, and even mammoths. It is likely that these hunter-gatherer groups consisted of only about forty people—living close to water sources and working collaboratively to trap and kill game using short spears. By about 8,000 years ago, the climate on the plateau had warmed and dried further, leading to changes in the profile of fauna that were adapted to this new niche. Bison still provided a source of food, but the staple diet was likely to have been smaller animals such as rabbits.[2] This was supplemented by roots, seeds, and fruits, collected in simple baskets and sometimes ground between stones or dried with hot coals for storage.

Like the first-nation Australians discussed in Chapter 1, these hunter-gatherer groups did not really construct a niche; instead they followed their game food sources and the seasonal changes in plants, migrating to the highlands in summer and the lowlands in winter.

From about 1500 BCE, corn (maize) and squash (pumpkin varieties) were growing on the plateau. These plants originated farther south in Mexico and perhaps arrived through trade. It appears that the people living on the plateau did not cultivate these crops intensively, instead planting seeds at particular sites as they moved, returning to harvest them at the appropriate time during their seasonal migration cycle. These hunter-gatherers did not settle to become emergent agriculturists until about 200 BCE.

These hunter-gatherer groups were not the only ancestors to invent agriculture. In fact, settlement and agriculture arose independently in at least ten places in the world. These places were so geographically separated, across several continents, that it is impossible to argue that this was the result of communication or a process of cultural diffusion of ideas between these populations. The earliest agriculture was probably in Mesopotamia about 10,000 years ago, and the last to develop was in New Guinea perhaps only a few hundred years ago.[3]

The development of agriculture enabled an increase in the rate of population growth. Agriculture can supply food for a population at least fifty times greater than the number of hunter-gatherers occupying the same area of land, assuming the land is arable or capable of supporting domesticated animals bred for food. World population increased perhaps fifty-fold from about 6 million in 10,000 BCE to about 300 million by 1000 BCE. The extent to which increasing food production permitted an increase in the population, or vice versa, is not known.[4] But as we will see in Chapter 6, larger conglomerations of people not only create greater demand for food but also can drive the creation of new technologies to produce it. In addition, intensive cultivation and the stratification of social roles permits a division of labor, which speeds up technological progress through "learning by doing" and leads to efficiency savings in time and energy.

The selection of varieties of plants for cultivation marks the beginning of artificial selection by breeding, a process that continues today, using further technological ingenuity in the form of genetically modified and gene-edited crops. Even during the early stages of agriculture, we actively selected plants to breed, such as types of cereals which gave a higher grain yield than others. As discussed in Chapter 1, Darwin's interest in such plant and animal breeding by artificial selection—selection by the breeder with intent—led him to the concept of natural selection. What evolution took many millions of years to accomplish, humans tinkered with over a few thousand years and, with new technologies such as genetic modification, can now accomplish valuable and desired plant evolution in less than a decade.

The selection of crops with a higher yield not only provided a more effective and efficient energy source for humans, it also provided feed for domesticated animals. These would initially have been wild species held in enclosures to remove the need to hunt them, but again our innovative skills soon turned to breeding selected variants with docile behavior, faster growth, and better meat or milk yield. If we consider the entire human food chain in the simplest terms, we see that it starts with the capture of energy from sunlight by photosynthesis in plants and phytoplankton. These are then consumed by animals, birds, or fish, which we then kill and eat. It has been calculated that humans now utilize perhaps a quarter or more of the energy trapped by photosynthesis on the planet—a consequence of how we have modified our environment.[5]

Staying Put

Groups that depended on agriculture for their survival were able to remain settled in one place. Formerly hunters and gatherers, humans now began to plant, tend, and harvest crops. They created a safe place to store seeds and produce, and guarded crops from scavenging animals, birds, and rival groups. For example, the basket-making people of the Colorado Plateau traded their baskets for pottery made by the Mimbres people to the south, instead of making their own. Trade thus becomes an important element of coexistence and the division of effort (and ultimately leads to many other innovations that we now see in political, financial, and economic systems). The accumulation of possessions such as pots and tools makes a nomadic way of life more arduous, so it is usually not until settlement occurs that possessions start to accumulate.

Other technical innovations also occurred on the Colorado Plateau. The development of bows and arrows, atlatls and slingshots, which are more accurate and lethal than throwing-spears, made hunting easier and less time-consuming. The populations of the plateau settlements grew, as did those of surrounding groups—the Fremont in what is now Utah, the Sinagua in Arizona, the Hohokam in the Sonoran Desert, and the Mogollon in New Mexico.

The ancestral Puebloan culture achieved its peak in what is now called the Four Corners area, where the present states of Colorado, New Mexico, Arizona, and Utah meet. As documented by the archaeological sites of Mesa Verde, in the sixth through the thirteenth centuries sophisticated settlements were developed, and the complex pit houses were developed into superstructures. Tasks became divided between agriculturists, potters or basket makers, builders, hunters, those who maintained the water supply system, and so on. Such specialization inevitably leads to benefits as well as conflicts. The exchange of goods and services requires cooperation and a sense of trust. But specialization also leads to hierarchies and social stratification. In addition to the hierarchies within the community, the distinction between those living in these ancestral Puebloan communities as intensivist members of the simple economy, and other groups living on their margins and posing a threat, led to a "them / us" or "insider / outsider" culture.

Such social supremacism, larger population groups, and potential conflict may have led to the construction of the larger, seemingly fortified Great House structures in Chaco Canyon, which appears to have been a major site for convergence and religious ceremonies until about 1120 CE. In Mesa Verde a new phase of niche modification commenced in about 1200 CE, with the construction of the extraordinary cliff dwellings with their many rooms and complex compartmentalization, set into the rock at an altitude of about 7,200 feet. The function of the hundreds of rooms, some windowless, is not known: perhaps they offered protection from winter cold, or were storerooms for harvested crops. Archaeological exploration here has also revealed thousands of artifacts of beautiful turquoise and shell ornaments, fine pottery, and musical instruments, suggesting that this was the site of a high-status culture. This led to examination of the patterns of DNA in the human remains at the site. Scientists can now extract DNA from bones in human remains, even from bones much older than those from the Colorado Plateau. The DNA is all that is left of the cells in the bone marrow that generated blood cells. DNA in all cells is present in two forms: nuclear DNA, 50 percent of which we inherit from each of our parents; and mitochondrial DNA, which we inherit entirely from our mothers in the cytoplasm of the fertilized egg. At the Chaco Canyon site, it turned out that a high proportion of the individuals buried there had the same mitochondrial DNA, so there may have been a matriarchal dynasty with the equivalent of a queen passing her mitochondrial DNA to her children and thence to a whole culture of descendants.[6] Such research suggests that the population was held together by a hierarchical structure of authority, at least until there was some massive social upheaval.

Group Rules

If cohesive groups are more able to survive and thrive, this might be considered a form of natural selection—acting at the level of the group rather than the individual. As we discussed in Chapter 4, to optimize its chances of survival, a group would need to have binding rules and a strong sense of identity, and be willing to punish members of the group who did not pull their weight.[7] The punishment might be social or physical—expulsion, imprisonment or death; or it might be spiritual—excommunication or condemnation to a life in hell in an afterlife. Modern gangs can be seen as a form of in-group. A gang is characterized by strong rules and rituals and severe punishment for those who break those rules; but the gang provides identity and a sense of belonging that is often valued by its members. In Chapter 4 we discussed the debates over levels of selection and whether it occurred at the level of the individual within a group or for the group as a whole, with its particular collective characteristics. The latter is sometimes called *cultural group selection*—the concept that selection can occur at the level of the cultural group and based on the survival and fitness advantages the group's cultural attributes provide.[8]

Settlement Challenges

During human evolution, some genetic changes apparently conferred advantages to primitive pastoralists relatively rapidly. As discussed in Chapter 4, one of the best-known examples is a mutation in the gene for lactase, the enzyme that allows us to digest the milk sugar lactose.

Once cities and towns formed, health risks started to grow. Infection associated with poor urban hygiene and epidemics were common. The impact of our niche modification behaviors on our health can be seen in changes in height. Archaeological studies show that height started to fall when agriculture emerged, because of the increased risk of disease, and fell further with urbanization. Height started to rise again only when innovative public health systems started to be developed in the nineteenth century.[9]

Historically, we can trace the origin of many diseases that afflict us today back to this period of societal innovation. Settlements led to living in close proximity not only with other humans but with domesticated animals and with pests such as small birds and rodents that fed on food scraps and stored crops. This led to novel opportunities for diseases to pass from animals to humans. One of the clearest examples is influenza. The flu virus is likely to have originated in pigs and poultry, a transmission mode that we still see today. The virus is highly mutable, and this allows it to escape destruction by the immune system. Occasionally a new strain emerges, such as H5N1 in chickens. As H5N1 spreads through the chicken population, there is a chance that it may infect humans who come into contact with the domesticated birds. Any human suffering from this strain of flu will have caught it from birds, as it is bird rather than human flu. However, the more humans are affected, the greater the risk that a single virus in one human will undergo another change, either undergoing a spontaneous mutation or obtaining some genetic material from an already transmissible human flu virus that will make it transmissible between humans. The result is a potential epidemic, and its spread will depend on the degree of contact.

The first modern example of such a flu epidemic, in 1918–1919, infected about 500 million people, about one-third of the global population at the time, and killed 20 to 50 million people. It may have originated in birds, and it may have first taken hold in the U.S. training and supply camps for soldiers in the First World War. Whatever its origin, the greater movement of people around the globe at the time of the First World War and afterward, coupled with close living quarters at camps and in transport, likely fueled the spread of the epidemic. It spread in two waves, the second even more virulent than the first, which together killed ten times more people than did the war itself. Human ingenuity in devising ways of fighting and killing people on a mass scale—including poison gas, tanks, aircraft, and more powerful guns—also resulted in this viral epidemic leading to global disaster.

Guns and germs are only some of the reasons societies struggle and may fail.[10] The population of the Mesa Verde area declined along with its agricultural output between 900 and about 1100 CE. By 1300 CE, the site had been abandoned, apparently rather rapidly. This may have been the result of the drought that started in 1276 and lasted twenty-five years, exacerbated by increasing conflict with neighboring groups. The challenge of securing defenses, along with providing sufficient food and a secure source of water, had apparently become too great. The shift from constructing a niche to modifying it to cope with changing circumstances requires technological innovations, whether developed within or absorbed from beyond the group, and makes great demands on any society. A group's survival is dependent on its ability to keep up the pace of such innovation.

Novelty and Nostalgia

There are several theories about the drivers of social change at critical times throughout history. One theory suggests that a critical factor is the exposure of the population to novelty, usually from another population group.[11] Novel culture may present a threat or an opportunity, and contact with it may occur for some time before it produces a major cultural shift. This is illustrated by the response of the Puebloans to the twenty-five-year drought that led to their sudden migration to the Tewa Basin. The Jemez Mountains in this region are a good source of obsidian, a glassy volcanic rock that can be fashioned into razor-sharp implements, especially arrow points, and which is far superior to the sandstone of the Mesa Verde region. In the period up to the abandonment of the Mesa Verde settlements, the number of obsidian arrow points found at Mesa Verde by archaeologists increases substantially. It seems that scouts returning to the Mesa Verde settlements or traders brought these points from the Tewa Basin. Increasing reports of life among the Tewa Basin dwellers would have been conveyed to the Mesa Verde residents, who were struggling under difficult conditions, until eventually a critical point was reached where a mass migration occurred.[12]

Given the promise of better hunting tools and water supply in the Tewa Basin, migration became attractive. When opportunities to improve their current niche literally dried up, shifting to another niche, albeit one already occupied by other groups, was the best option. We can see that the migration was deliberate from the evidence of fire damage in the abandoned pit buildings (kivas) in the Mesa Verde settlements, perhaps a sign of ritual abandonment by people who left knowing that they would never return. Further evidence comes from the more desperate innovations of the few people who for some reason remained. The kivas are increasingly grouped together into defensive structures, and there is evidence of violent struggles among them. By contrast, in the Tewa Basin there is evidence of cultural assimilation between migrants and residents, as seen, for instance, in changes in pottery decoration styles. Societies have many ways to preserve cultural values and practices from the past, to ensure social stability and collective cooperation. Probably one of the strongest is religion.

Religion

Humans are essentially social animals. When humans moved out of the Neolithic into settlements they formed larger and more complex social groups with cultural traditions, ideologies, and rules. Increasingly, the groups were no longer homogeneous but much more diverse. As groups expanded or merged with other groups on shared land or settlements, rituals that bonded them became more important to create cohesion and boundaries. As group identity became stronger, there emerged the concepts of "insiders" and "outsiders." Groups of people who have common beliefs and interests are more likely to unite against a common enemy or to avoid trouble all together. And survival to a large extent depended on being part of a group. As towns and cities formed, cohesion of beliefs and respect for the local elite became critical for both the ruler and the ruled. And so institutionalized religion and political power became entangled. Even today the Queen of England is head of the Church of England, the pope is treated as a head of the Catholic State, and Iran's supreme ruler is its religious leader.

The complexity of society that came with settlement led to the formalization of political and religious systems. Religion may have evolved as a social response to the natural world, where despite the comfortable life provided by good harvests and healthy domesticated animals, there was always the risk of natural disasters such as droughts, floods, or disease. We all tend to be superstitious about the "accidents" in life that we cannot explain or predict, and it is easy to envisage our early ancestors as having needed to attribute these to an external, supernatural force that required propitiation. To this day, many people want to blame someone else for their misfortunes. For example, after the earthquake in Lisbon on Saturday, November 1, 1755, there was widespread consternation in Europe. The earthquake had occurred on All Saints Day, at 9:40 a.m. local time, along with fires and a tsunami (likely caused by a massive underwater landslide triggered by the quake). It almost destroyed Lisbon and killed perhaps 100,000 people. At the moment of the quake, many of Lisbon's residents were in church. So, even though other European populations were not directly affected by the quake, many saw it as an act of divine retribution, a punishment for immorality in European societies as a whole.

In a settled community where tasks are stratified, there can emerge shamans or other religious leaders with various kinds of power and authority over the population. The eminent evolutionary biologist David Sloan Wilson and others argue that organized religion provides a way to control freeloaders within a large society—for example, by fostering the belief that such behavior will bring retribution to the population as a whole.[13] A common feature of many religions is that they posit a supernatural entity or force that has a population's survival at heart and therefore rewards altruistic behavior and generalized reciprocity and punishes selfish behavior.

Many religions also promote belief in the significance and importance of rituals. A ritual might take the form of specific acts at certain places to ensure continuity of control over members of the population regarding critical activities, such as crop harvest or mate selection. Rituals can signal membership in a group and create social cohesion.[14] A sacrifice to the supernatural (such as fasting or, in the past, even human sacrifice) can provide evidence of loyalty to the group. Rituals may also have a deeper origin in the checking behavior which is necessary to avoid harm under certain circumstances. For example, the washing and cooking of certain foodstuffs in ritualized ways, or avoidance of certain foods, might have evolved as behaviors that reduce risk of exposure to toxins or pathogens. Pilgrimages to certain sacred sites might be helpful in retaining territorial boundaries or endorsing the role of a ruling elite. As societies became more complex, belief in supernatural beings became a way of giving greater authority to rulers who assumed the role of their agents. Thus, organized religions became instruments of the proto-State and continue today to be central to the constitution, legal structure, and identity of many nations.[15]

As social groups grew in size, they became more stratified by profession, wealth, and other forms of status. Strong mechanisms can be used to maintain this stratification, as was characteristic of feudal and other pre-Enlightenment forms of governance in Europe. Echoes of this persist in the sense of class difference that pervades many societies and, in its most extreme form, in caste system, as in India, or in the religious entitlements of some sects in some religions—for instance, there are restrictions on mate selection for the Cohanim, the hereditary priestly caste of Orthodox Jews.

Keeping Up

In the niche-modifying processes of developing agriculture and settlements, we can see a cultural evolutionary path based on cumulative technological advancements that had considerable immediate advantages. Over the long term, though, those innovations had negative impacts, in response to which people had to invent new technologies (sewage disposal, antibiotics, vaccines, disinfectants) and social structures (public health systems, the military, monarchies, universities, priesthoods, democracies) to cope with the environmental changes we had created. In Chapter 6 we explore this phenomenon and how, as persistent niche modifiers, we have to keep inventing new technologies to cope with the consequences of our previous modifications—we have to keep on trying to change nature. This raises the question: Can we keep on doing so indefinitely? More than half of the world's people now live in urban environments, and their number increases every day, so to answer our question we must turn our attention to how cities help us to change nature. But, as we are beginning to see from our previous discussions, we also have to ask whether the choices we've made are changing our nature as well. At what price, our ingenuity?

6 | CITIES

Ingarsby, a former medieval English village, now in ruins, takes its name from a Danish invader or settler and means Ingwar's village. The village would have been established toward the end of the ninth century, although there would have been some settlement there before that time. It is located on a hillside in Leicestershire, a part of England about as far as one can get from the sea. But on hillsides such as this, deposits from the last Ice Age provide well-drained fertile soils for cultivation and pasture. During the Norman period, following the invasion of 1066, Ingarsby was large enough to be recorded in the Domesday Book of 1086. Slowly but surely the landscape was cleared of woodland to allow more cultivation, and the population grew. It must have been backbreaking work with nothing more for tools than simple iron implements and cattle to pull carts, but a few acres of cultivation nonetheless were added each year.

By the thirteenth century, Leicestershire was one of the most populated counties in England, and that population could not have increased much more without gaining access to more land. The territory of the village had to be guarded; the ditches surrounding Ingarsby, along with some wooden stockades, served this purpose. Not only was it essential to defend the grazing cattle and cultivated fields against raids from surrounding villages, but it was also important to be largely self-contained. Many villages at that time introduced innovations such as water mills that helped mill grain into flour. But some materials, such as iron, could only be obtained by trading.

The balance between agricultural production and population during this time, notably in this landlocked part of England, can be seen in terms of the "Malthusian Trap" as later expounded by Thomas Malthus in *An Essay on the Principle of Population* published in 1798.[1] Malthus argued that agricultural growth is essentially linear. For example, every new field added to those cultivated by a village will give a similar increase in crop production or number of cattle that can be grazed. In contrast, population growth is exponential; it grows by a process of approximate doubling over each unit of time, like the growth of a colony of bacteria. So a couple having two children who have survived adolescence can expect to have at least four grandchildren, eight great-grandchildren, sixteen great-great-grandchildren, and so on. And if they had more than two children, as was the norm at that time, population growth would be even faster. Having more children may have generated more potential labor for agricultural work, but these were also more mouths to feed. There was a real risk of starvation, especially in a year of bad harvest or if cattle died of disease or were stolen. The most vulnerable were the young and the elderly. Malthus argued that because of these effects, the size of a population would be limited by the resources available to it. If it grew slightly larger than this optimal size, food would become too scarce to support the additional number of people and the population would diminish again.[2]

Plague!

As it turned out, however, it was not a "Malthusian Trap" that challenged survival in Ingarsby. It was the plague, caused by *Yersinia pestis* spread from Europe by fleas arriving on rats aboard ships. The concept of germs did not exist at that time, and the plague was attributed to many competing and often supernatural causes, such as contaminated winds from Asia or lack of faith in God. The Black Death of the second half of the fourteenth century claimed the lives of 30 to 60 percent of the population in some parts of Europe. Contagion was greatest in the cities, due to close quarters, but transmission was also high in villages like Ingarsby.

A Village Dies

Although the population increased again following recovery from the plague, a new innovation sealed Ingarsby's fate—a simple change to the layout of the landscape drastically altered the economy. The growing market for wool meant that sheep farming was now becoming highly profitable.[3] Large tracts of land were enclosed, abolishing strip-farming by peasants. Within the space of a few months, Ingarsby was converted into a large sheep and cattle ranch of about 1,100 acres. Because there was no longer any need for peasant farmers—one shepherd and his dog could now look after this estate—the village was abandoned permanently in 1469. What was once the main street is now simply a depression in a field, with mounds separated by shallow ditches on either side going down the hill where the mediaeval cottages stood and lanes between them once ran. The character of the Leicestershire countryside, and the lives of its inhabitants, had changed forever. The property around Ingarsby was the most valuable grange in Leicestershire when it was sold at the time of the Protestant Reformation in 1540. A new breed of landowner was produced almost overnight. These landowners continued to profit from raising sheep and cattle in high Leicestershire, retaining the large fields from the earlier enclosures and reestablishing some woodland only to cater for their hobbies like fox hunting.

The economic and accompanying social forces that led to the demise of Ingarsby are far from unique. They are part of a wider pattern of cultural evolution that affects not just villages and towns but entire countries. In their book *Why Nations Fail,* Daron Acemoglu and James Robinson analyze the histories of a range of cultures in terms of the long-term harmful effects of extractive economic systems, where energy is put into maintaining the economic status quo for an elite, at the expense of the poor.[4] To avoid the almost inevitable resulting failure of the economy and society, they argue, investment in new technologies is vital. As we will see in Chapter 7, this can generate a middle class and the roots of a more democratic system—arguably progressive steps in cultural evolution.

Fenced In

The massive changes in England's rural environment caused by the building of fences and hedgerows were progressive, culminating in the parliamentary Enclosure Act of 1773, which gave landowners the right to enclose their land and deny public right of access. In theory, such enclosure was to be preceded by public consultation, but this was often hasty and clandestine, like many political consultations of today, so that enclosures frequently evicted small farmers who could not prove their legal right to remain, along with villagers and any poor peasants who happened to be eking out an existence on a pocket of land.

These technological innovations, while now seemingly rather simple, produced immediate benefits for some people, as well as longer-term consequences. When previously farmers were working the smaller plots, poor drainage or inadequate weeding by one farmer could have negative consequences for the productivity of his neighbors. Making larger fields from a series of smaller plots resulted in less wastage of land and more-consistent output. Larger fields allowed the use of machinery such as seed drills, which could be shared by large landowners, less risk of the spread of disease in cattle, and possible new techniques such as crop rotation. Such "new husbandry" had been pioneered in the countries of Northern Europe hundreds of years earlier. More significantly, however, the greater efficiency of these large farms meant that they needed fewer laborers. Within a generation or so, families who were subsistence farmers had to change their lives. Where before they could grow most of their food, now they needed money to buy it. This meant taking up paid employment. The larger estate farms were very profitable, especially through wool production, but generated rural unemployment. Where could these displaced families go in order to survive? To urban environments, where new possibilities were developing.

The Industrial Revolution

The changes in agriculture in England following the Enclosure Act sowed the seeds of the Industrial Revolution and the shifting politics that followed. Karl Marx published the first volume of his magnum opus, *Das Kapital,* in 1867. The book did not sell well initially, even though Marx's ideas were later to become extremely influential. Marx learned much from discussions with his friend and collaborator Friedrich Engels, himself a manager of a textile factory in Manchester and whose own book *The Conditions of the Working Class in England* had been published in 1845. Engels's perception was that the "science of enrichment," applied systematically to the workers in the new industrial urban areas made them no longer artisans or peasants able to subsist on a simple rural economy, but wage-slaves with no property or rights. In *Das Kapital,* Marx extended this idea to argue that the owners of the mills and factories purchased the workers' ability to labor, not their labor itself. By this means, they could extend working hours and focus on productivity levels to generate "surplus value" in terms of production of goods, and thus profit at the expense of the laborer. The principle that capitalism depends on growth through profit and reinvestment was thus established.

Marx pondered the effects of technology and mechanization on the workforce in his notes, speculating that they could lead to a "capitalist crisis."[5] If competition between producers in England and overseas brought prices of their products down, then profits would fall, challenging the security of workers' employment itself. Mechanical innovations such as looms to make cloth in the mills might replace labor, potentially saving money, but by increasing production efficiency, they might also bring profits down further. In a utopian dream it might be imagined that machines would take over the drudgery parts of work, giving more time for leisure and more interesting work. These are very modern considerations, to which we shall return in Chapter 8. Marx was also interested in anthropology, especially the work of the American anthropologist Lewis Henry Morgan, whom we encountered in Chapter 4 and who had developed the concept of cultural evolution.[6]

Expansion

The Industrial Revolution is often viewed as having brought the first wave of modern innovation. As populations increased and new skills were developed, many based on new technologies, specialization became more narrowed. The division of labor, along with investment from new markets, led to increased productivity. Agricultural productivity increased about 350 percent between 1700 and 1850.[7] This fueled the new industries and gave wages to at least some of the increasing number of displaced families from the rural sector. The increase in food production supported England's growing population, averting the kind of Malthusian crisis experienced by many other countries, such as India and China, in the nineteenth century and reoccurring today in parts of Africa.

The primary industries in Leicestershire had been hosiery and footwear. Hosiery weaving, using locally produced wool, had rapidly become a cottage industry. At the end of the 1700s, three-quarters of Leicestershire families lived in rural communities. One hundred years later, two-thirds of Leicestershire families lived in urban areas, half of them in the city of Leicester itself, which had increased in size thirteen-fold in a century.

Leicester's size and population density were growing fast, but not in such dire ways as in its neighboring city Nottingham. Unlike Leicester, Nottingham was encircled by fields of common pasture land that could not be developed for housing, and the housing within the city itself became much denser. Nineteenth-century builders in Leicester could build out from the city center, constructing new streets of small terraced houses; in Nottingham the increasing numbers of people who worked in the lace and hosiery industries were crammed into tenements, generating some of the worst slums and health conditions in Europe.

The real power behind the expansion of industry in cities such as Leicester was, literally, sources of power. In early human settlements and early agriculture in many parts of the world, the innovation of using animals to draw plows or thresh grain was critical. In the eighteenth and nineteenth centuries, industry was driven by innovations using water power to drive spinning machines, and steam engines to pump water, drive engines, and move goods and people. These sources of power drove a second wave of innovation.

The competitive economic environment of Europe increased such innovation, as did the demand and opportunities created by colonization and the development of the forerunners of global corporations, such as the East India Company. The third wave of innovation consisted of new industries—for example, dye and fertilizer manufacture in Europe—along with the development of electric power.

Industrialization and increasing urbanization brought with them enormous social changes, which formed the basis of much of what we now see as social responsibility. For example, as hosiery knitting shifted from family-based units in rural communities to small shops and factories, a wage system became necessary. But workers in smaller shops were sometimes paid partly with a share of the goods they produced, a system that allowed the managers to pay workers little and keep them indebted. It was not until after the 1844 report of a Royal Commission into the conditions of framework knitters that the concept of payment of a "working wage" arose, but its level has been disputed in various ways ever since and remains at the heart of much contemporary political debate.

Because factory workers may have felt less loyalty to their employer, and put in less work than they would for a family business, it was important to create some form of control of attendance, productivity, and standards of output. The behavioral codes of the human social animal were extended to include a sense of commitment, of duty perhaps, to a firm and the wider economy. This regulation of behavior extended beyond kin, and beyond a Dunbar number of contacts or a religious or political group, to the economically relevant group. And so life became more complicated. Not all these rules and expectations necessarily aligned—think of the modern and very personal debates we have over work / life balance or the confusion of politics, economics, and religion seen in many contemporary countries.

Megacities

Cities are complex environments.[8] They attract and retain a wide and disparate range of individuals and groups, and today their populations can be extremely large. Megacities, with populations in excess of 10 million people, have grown extremely rapidly in recent years. In China, within a single generation, there has been a massive movement of the population from rural to urban environments—1.3 billion Chinese people now live in cities. Dhaka in Bangladesh has an estimated population of 15.7 million and a population density of 43,500 per square kilometer. Compare this to 2,400 people per square kilometer in Los Angeles and 4,400 people per square kilometer in Tokyo, or to recently expanded cities in China—especially Guangzhou-Foshan, which has a slightly higher population density at 6,000 people per square kilometer but its overall population is 20.5 million.

There have been a number of detailed analyses of how various aspects of our lives scale with increasing sizes of cities.[9] For example, in studies of contemporary cities in the United States and China, the application of ingenuity in the form of developing patents or employment in research and design activities increases faster than population size, showing how city life increases the range of contacts and favors creativity. Increased contact, however, can also have negative consequences, and crime and HIV levels also increase faster than population size. There are also aspects of city size that scale more slowly with population growth, such as the number of gas stations and lengths of road. These may represent valuable economies of scale.

The close proximity of different social groups, often viewed as "classes," within a city can generate tensions and problems. This is part of a much bigger question about the advantages and disadvantages of encouraging or allowing groups to live in specific areas within a city. Such areas, which increasingly have ethnic identities, at least in Europe, North America, and Australasia—if not outright labels, such as "China Town"—have sometimes been established deliberately, rather than growing spontaneously, and so have a range of intended and unintended consequences, as the sociologist Robert Merton noted more generally.[10] Zonation can lead to a ghetto effect, where property prices fall, unemployment rises, and young people in particular feel disenfranchised and can revert to crime and other antisocial activities. On the other hand, distinctive areas within cities can have the advantages of bringing together in close proximity people with similar talents, interests, and skills, enabling easier exchange of ideas, and resulting in greater creativity and innovation. Innovation zones—concentrations of specialist professions such as artists' studios, restaurants, tailors, gamers—are becoming particularly fashionable. It might be appropriate to dismantle some urban zones, in order to remove stigma and disadvantage for ghettoized ethnic groups, but reduced clustering can also reduce the cultural diversity and creativity of a city as a whole. This suggests the need for a more sophisticated and informed approach to city governance and management.

An even more contentious issue is urban poverty. One of Edward Glaeser's emphases in *Triumph of the City* is that cities attract poor people. For Mumbai, India, today, perhaps as for rural Leicestershire at the time of the Industrial Revolution, moving from intense rural poverty to a different type of poverty in an urban setting may have been lifesaving for some individuals and families. However poor, many urban settings offer some employment opportunities, social services (although often rudimentary by the standards of nearby affluent areas), and opportunities in terms of exchange of ideas and so on. The highly problematic issue is that, while it is true that social mobility is predominantly one-way in an upward direction, it is also true that the poverty trap is so deep for many families that they will not escape for generations, if ever. Transgenerational disadvantage is one of the biggest current policy challenges for liberal democracies.

An Unhealthy Environment

Some of the best-known effects of city life on health are the increases in the risk and prevalence of infectious diseases. Recall the transmission of the plague. High-density habitation and poor sanitation provide excellent conditions for transfer of a wide range of pathogens among the population. Diseases such as diphtheria, measles, influenza, and smallpox are spread by coughing or direct contact, and historically they caused high mortality in cities of what is now the developed world. They still do in urban areas of low- to middle-income countries. Many other infections, such as typhoid, typhus, and cholera, were and are still spread through food or water contaminated with human feces.

Until relatively recently there were no effective treatments for most of these diseases. Technological means to prevent their spread are crucial—including even basic interventions like John Snow's simple act of removing the handle of a public water pump in London in 1854.[11] Snow halted the spread of a cholera epidemic simply by preventing access to a contaminated water supply.

A second line of defense is personal hygiene. In the mid-nineteenth century, the provision of soap and water for washing and the development of sewer systems were interventions that dramatically improved life expectancy for city dwellers. Such notions of personal hygiene were controversial when first introduced. In 1847 the Viennese physician Ignaz Semmelweis demonstrated that hand-washing and disinfecting with chlorine reduced the risk of often-fatal infection in women at childbirth—lowering the fatalities from more than ten in a hundred to less than one in a hundred.[12] His findings, based on astute observation, were ridiculed and ignored by the medical profession. It was not until 1884, when Robert Koch developed his postulates, following the groundbreaking work of Louis Pasteur in the 1850s, clearly showing the role of germs in the etiology of infectious diseases such as cholera and tuberculosis, that the importance of handwashing was established beyond question.[13] Without the means to identify the pathogen that caused a disease, or to demonstrate the mode of its transmission from an infected person to an uninfected person, innovative steps to control the spread of infection were met with bemusement or even disdain. Semmelweis died in an asylum—ironically, probably from infection.

Sometimes preventative innovations against infectious disease can be developed even without understanding the underlying processes. In 1796, in just such a way, Edward Jenner developed a method of vaccination against smallpox. Cowpox was produced by a virus that was less virulent than, but immunologically similar to, the virus that caused smallpox. Realizing that cowpox was sometimes contracted by women who milked cows, and that these milkmaids were known to be protected from smallpox, although it was not known why, Jenner used fluids from cowpox blisters to successfully inoculate against smallpox. Pasteur and then many others, including in more recent times Jonas Salk and Albert Sabin, built on Jenner's creative inoculation procedure, and now many bacterial and viral diseases are kept under control—and are almost forgotten, at least for the present in developed countries—because of immunization and vaccination.

The first attempt to develop a chemical to kill bacteria was by the Nobel Prize–winning German chemist, Koch's friend Paul Ehrlich, who in 1907 developed an arsenic-containing chemical to treat syphilis. This was followed by the sulfonamides developed in the

1930s, which are still in common use in evolved form, followed in the 1940s by the large-scale production of the first true antibiotic, penicillin, which had been serendipitously discovered by Alexander Fleming in 1928. The sulfonamides were chemicals initially derived from the dye industry in Germany after extensive research. Penicillin, in contrast, was a biological substance produced by a fungus, and it presented an entirely new approach to treating infections based on the natural defense mechanism of the mold. In a way, it set one organism against another. These innovations offered great hope for the treatment and control of infections, especially as research in the United States led to the discovery in 1943 of streptomycin. This drug was derived from a soil bacterium and was effective against tuberculosis, which had proved intractable using previous antibiotics. Within a short period of time the isolation hospitals and the ineffective treatments for tuberculosis were abandoned. Streptomycin is also effective against the plague bacterium *Yersinia pestis,* so life would have been very different in fourteenth-century Leicestershire and elsewhere had this innovation been discovered 600 years earlier, at the time of the Black Death.

Our ingenuity led to many triumphs over bacterial disease through the discovery of antibiotics. It was soon discovered, though, that bacteria reproduce very rapidly, and thus have the ability to mutate rapidly—which led to the emergence of bacterial resistance, and every time we found a new antibiotic and started to use it—often inappropriately—resistance was soon to emerge. It is an arms race that continues to this day. But we are now in trouble, as the pace of discovery of new antibiotics has stalled and we face the growing problem of antimicrobial resistance. Is there a limit to our ingenuity that in the end allows nature to bite back? This is an important question, to which we shall return.

A Sense of Well-Being

Studies in developed countries suggest that more than 20 percent of the population at any one time are in need of psychological or psychiatric support because their mental state is not optimal.[14] Indeed, urbanization appears to be linked to increased risks of schizophrenia and other forms of non-affective psychosis, even in children.[15] The data are sparse, but the general sense is that the need for mental health support in urban populations is rising. There are many reasons this might be the case, but they all point to the need to understand our well-being from an evolutionary perspective.

As we discussed in Chapter 5, we evolved to do best in social terms in relatively small groups with clear rules and identifying factors that defined our groups. But the modern city is not a series of small, clustered urban villages and neighborhoods. Instead, urban environments have very diffuse social structures—residents live in close proximity, in communities much larger than Dunbar's optimal 150 people—and such an environment does not give its residents a firm sense of identity. Today there has been a marked shift in the perceptions and realities of societal cohesion, even in traditional democracies, trends fueled by changing media and social media that are promoting forms of tribalism and conflict. This is particularly reflected in the politics of populism, the increasingly partisan and tribal nature of politics, and increases in racism and other forms of fractious societal division.[16] Urbanization has played a large part in this. The rise of gangs, for instance, can be seen as an evolutionarily logical attempt by those who feel confused, disaffected, or disadvantaged to create a strong group with rules—rules that might look antisocial to those outside the group, but that feel coherent and induce loyalty and identity for those within the group.

Secondly, family structures have changed. Gradually, following the Industrial Revolution, the concept of the extended family living together evaporated in most modern urban societies. The psychological support inherent in extended family structures has been replaced by smaller family groups and less-formal relationships. Some evolutionary psychologists see the breakdown of traditional kin relationships as very important. For example, Canadian scholars Martin Daly and Margo Wilson controversially suggested a possible evolutionary argument for stepfathers being more likely than genetic fathers to commit child abuse or murder, along lines similar to the male lion taking over a pride and killing suckling pups fathered by a prior male.[17]

Our brains are highly responsive to new environments, and even in crowded urban environments, advertising and social media appear to be able to alter our mental states. Digital technologies affect our identity, beliefs, and relationships with each other and the society we live in.[18] Today not only do most people live in physically close proximity—our minds are also becoming more crowded. Most people now also "live" online, no matter where they are physically. In these new technological environments, the result of a new wave of ingenuity, humans are evolving again, adapting to this digital world of human making. We will explore how we are changing our nature within that world in Chapter 7.

7 | ONLINE

When we were writing this book, we were often 10,000 miles apart. But despite this, we were almost continually interacting, working together, discussing, analyzing, and writing. In addition, we had jointly built up an international research consortium spanning Auckland, Singapore, and Southampton, UK, and we needed to get to grips with new data from our labs that seemed to have exciting potential. But was the analysis right, was the validation complete, were the statistics sound, were the data sufficiently conclusive to draw strong conclusions, and, if they were, what were the implications? Or were the data incomplete, so that we should resist the temptation to speculate about them? All of these questions were going through our minds as we interacted—online. Like most scientists today, we see this as normal, but such a way of working is a very recent phenomenon. We have seen it develop over the course of our careers. Let's look back.

Magazines and the Scientific Disease

When we started our scientific careers in the 1970s, every week a little magazine called *Current Contents* would appear. It listed the contents of all the life sciences research journals—perhaps about 150 of them. We would look through it, identify publications of interest, and send off a postcard to an author asking for a copy of the paper, which might arrive in the mail two to six weeks later, if we were lucky. We relied on this process to know what was going on, along with evenings spent in libraries, engrossed in the rather small number of scientific journals in which we might find papers we needed to read, perhaps twenty at most. This was supported by personal interactions with other researchers in the corridors of the scientific meetings, which we might be able to attend a couple of times a year. When and if we received a copy of a paper, it might refer to another paper published in a more obscure journal or book that the library did not possess. Then it might take months, if ever, to track it down. It once took Peter over a year to obtain a copy of a paper that was important to his research at that time but had been published in a low-circulation Eastern European veterinary journal.

Then, in the 1970s, the National Institutes of Health, in the United States, started to produce an annual book listing all the papers it was aware of that were relevant to medical research, indexed by both keyword and author. This massively increased our access to information. We soon came to rely on that series of very large volumes to know what was going on in our field. But if the field was new, and our field of fetal and neonatal physiology was indeed new at the time, keywords were not always very helpful. The only salvation was that we thought we knew all the researchers engaged in the field. But when a new group formed, say in Japan, it could take ages for us to identify the authors, read their work, and start interacting with them. We worked like this through the 1980s, relying on face-to-face meetings and scientific gossip at meetings, and struggling with a tedious and incomplete way of keeping abreast of scientific progress. And the struggle became much more complex as the size of the academic enterprise grew rapidly in the 1980s and 1990s, when governments started expanding their investment in higher education and research, recognizing the importance of science to economic growth.

In the 1990s everything changed. Email, the internet, and the development of large online searchable databases of the scientific literature emerged. Science was transformed, as was its culture. But the change was not always for the best. As science spread on the internet, more journals emerged, not always with the same rigorous standards of the previously much more constrained scientific literature. Business models emerged where effectively almost anyone could pay to have their "science" published in journals of at best marginal quality. It became a challenge to separate good from bad, as science became submerged in an enterprise that by 2016 numbered about three million scientific papers a year in some 30,000 so-called academic journals.

But computation and digitalization, which had enabled this mass expansion, also meant that all sorts of statistics, whether meaningful or not, could be applied to the scientific enterprise. It was effectively now an industry—for many scientists, the primary reason to publish was not to participate in the pursuit of knowledge but to create a key performance indicator for their case for employment, promotion, status, tenure, and funding in the massively expanded academic sector. Scientists became recognized and ranked, not just by the perceived importance and broader impact of their actual work, but by how many papers they had published, and in which journals, and by how many times their papers were cited by others. Not all of this was bad, because it gave some indication of output and relevance. But combined with the massive expansion of the scientific enterprise and the emergence of "predatory" journals that only wanted profit and did not care about scientific quality, fundamental and unhealthy changes in the scientific system emerged.

This was the arrival of "bibliometric disease"—to borrow Oscar Wilde's phrase, "knowing the price of everything but the value of nothing." Universities and research institutes started using bibliometrics to evaluate scientific output and their researchers' achievements, sometimes even in place of effective staff development and performance reviews. Rather than looking at particular contributions and understanding the narrative of a scientist's work—Had it changed scientific understanding? How did it fit with others' contributions?—they simply used metrics related to publication. The scientific culture became "publish or perish." One had to publish in high-impact journals or fail to get funded, be promoted, or even stay employed.

Other kinds of contributions—those that have real impact—were submerged by this pseudo-objective culture. No matter how important they might be to society, contributions to public understanding or public policy were not rated as highly as "original" papers in "high-impact" journals. Some degree of recognition was given where publication fed commercial opportunities for a university, but even that played second fiddle to "citation" ratings, the number of times a particular paper was cited in other papers. Soon the disease spread to the ranking of whole universities. Even for universities that had a few superlative and world-leading researchers, what came to matter was only the digital ranking of a department and the institution as a whole.

This new culture of science increasingly focused on disciplines that could attract high ratings. Intellectual pursuit of knowledge was replaced to a significant extent by obsession with the short term: How many papers can be published, and in which high-impact journals, with this research program? The impact factor says nothing about the value of the paper. Even papers published in the journals with the highest impact may be rarely cited. But these rankings are now a significant part of the reward system in science. The cultural evolution of the academic science industry has been evolving to the point where it undermines itself. Only now are academies, funding bodies, governments, and international science organizations, such as the International Science Council, starting to think about how to address these unintended consequences of greater investment in science and the misapplication of incentives and metrics within the system.

Tables and Graphs

The interaction between society and evolving technology became much more apparent across the course of our own professional lives. For example, when we were high school, in the 1960s, we were introduced to logarithms. These are mathematical devices that make it much easier to multiply and divide large numbers. Much of science and technology involves large numbers—yet the processes of long division and multiplication by hand are tedious and sometimes practically impossible. Logarithms probably originated in Babylonia some 4,000 years ago. In the 1600s, European scientists rediscovered them and developed them into what we now know as logarithms. By expressing numbers as powers of other numbers (most commonly as a power of 10), it is possible to express any number simply, no matter how large or small.[1] The invention of logarithms led to the invention of the logarithmic scale. Instead of a graph of equally spaced numbers such as 10, 20, 30, 40 . . . , a logarithmic scale uses 1, 10, 100, 1,000, . . . This was an important breakthrough because many things in science show logarithmic relationships, so it enabled new understanding of scaling processes. For example to double or triple the effect of a drug on the body, we may have to increase its dose by applying logarithmic scaling. Slide rules were developed to display numbers on a logarithmic scale, so that multiplying or dividing the two numbers could be done without resorting to tables. By the 1960s, many more complex mathematical conversions, all expressed in logarithmic scales, were on slide rules, many of which had become quite complex.

Buzz Aldrin took his slide rule to the moon in 1969. In the same year Peter started his first scientific project. The work involved what then seemed like complex calculations; it would take half a day to conduct a simple statistical analysis involving perhaps 100 measures. But now there were mechanical calculators, which had their origin in the abacus. In the early 1600s, European scientists and inventors started to think about how machinery could be used to perform complex calculations. In 1642 the French mathematician Blaise Pascal solved the practicalities necessary for a clockwork mechanism to carry a number over (where, when we add 2 and 19, we carry a 1 over to the "tens" column).[2] But his spring-and-clockwork mechanism was not suitable for manufacture, and many mechanical improvements were made over the next 200 years. Mechanical calculators were first produced commercially in 1851. They were still basically adding and subtracting machines. By the 1960s these machines were at their high point—ten-digit machines were the norm, and they were electrically powered. But they were still basically adding and subtracting machines with clever but hidden subroutines for multiplication. The consequence was that any analysis was slow and simple, and the range of statistical methods possible was limited by the math involved.

Then in 1970 our world changed again. The first electronic calculator had appeared. It was the size of a typewriter and had only two memories—for one number each—but it could do some things in addition to multiplication, such as calculating a square root. Suddenly basic statistics became easier. And at about this time computers first became available for research use—but they took up a whole room. If we needed to do any complex calculations, we had to go to a special computing lab in our universities, where both data and the program to be run were entered on punched cards. Then we had to wait for an opportunity, usually after midnight, to run the analysis. When Mark returned from his medical expedition to Nigeria in 1971, what would now be considered a rather simple analysis of the data—

to relate patterns of contact with water in a rural environment with the level of the waterborne disease bilharzia—took several hours overnight on the university's PDP 1100 computer. The machine operated using radio valves, which generated a great deal of heat, so it had to be kept in a specially air-conditioned room. An early-morning trip into the computer lab to collect results was often met with disappointment as the machine had overheated and shut down.

The pace of innovation accelerated. By 1974 Peter had access to a primitive desktop computer, which had a whole 64 kilobytes of memory, although a trip to the giant computer housed on a different campus was still needed for any complex calculations. In 1982 we both got our first desktop computers, with a 5¼-inch floppy disk drive, and word processing became possible. No longer did a whole manuscript, thesis, or grant application have to be retyped to deal with a few simple mistakes or changes. And with increasing powerful computation in the hands of the individual, whole areas of science—environmental sciences, human sciences, biological sciences, planetary sciences—opened up to new forms of analysis. Complex systems like the global climate system could start to be analyzed in ways that led to important new understanding. Models of the function of the heart could be built, and the effects of drugs on interacting biochemical pathways could be calculated.

Vacuum Tubes and Silicon Chips

How was such rapid technological evolution possible? What allowed us to enter the digital age? This transformation would likely irreversibly change the human condition—but would this be for the better?

The mechanical calculator as we knew it had been essentially a simple commercial adding machine with modifications. In 1834

Charles Babbage had started to design a much more sophisticated machine in which data were entered through one process and a program for doing higher calculations through another.[3] His collaborator—and Lord Byron's only legitimate child—Ada Lovelace had produced what is thought of as the first algorithm. The machine was never actually built until recently, but conceptually it was the forerunner of much that was to follow. This was the basic architecture of most modern computers until the invention of supercomputers and quantum computing.

But the computer and the digital age depended on many more scientific discoveries and innovations. It required Faraday and others to make electricity useable.[4] It would require J. J. Thomson's great contributions to understanding the electron. It would require and take advantage of the gradual emergence of electronics, the development of the transistor and of computer languages. And it required the binary code—the heart of the digital age—which allows every number (or letter or other symbol) to be expressed as a series of 0s and 1s.[5]

In the 1930s the great British scientist Alan Turing conceived of a general approach to electromechanical computing relying on this concept, and in 1937 he built the first device. It was a series of electromechanical switches that could address sophisticated mathematical questions. In 1945 he worked on developing an electronically programmable machine. In 1948 the first such computer was built in Manchester, England. These early computers relied on vacuum tubes for their electronics. In the 1950s transistors took over—they were smaller and generated much less heat and were much more reliable. The computer age had arrived.

These were the first computers available for general use. IBM started to make such machines, the company very much wanting to forget that in 1943 its president Thomas Watson had announced, "I think there is a world market for maybe five computers." IBM's research park in Hampshire in England occupies an eighteenth-century stately home with grounds of a hundred acres, buildings to shame many a university, and a staff of over 1,500. It is the site where some of the world's most innovative software has been invented.

The age of the transistor did not last long. Soon there arrived silicon chips and the first desktop computers, handheld computers, and then the cell phone. The first of these phones were large and unwieldy, but they evolved rapidly. Parallel invention occurred in many companies. Cultural selection was truly under way in the competition between cell phone manufacturers—it was the survival of the fittest in terms of utility and fashion in the eyes of the consumer. And it was the selecting agent, the public, who determined which of these multiple approaches was to survive. BlackBerry featured a QWERTY keyboard with physical buttons to allow easier texting and email, and for a while that dominated the market, especially with business, but the screen was small. Apple recognized the value of a large touch screen with virtual buttons and most importantly the ubiquitous "app"—which has many characteristics but most of all it allowed us to do two things—to buy things (especially music) without entering credit card details every time, and to customize our cell phones for our own purposes. BlackBerry was soon in trouble—from which it never really recovered. Darwinian selection continues. Nokia, one of the great cell phone brands of a decade ago, has disappeared—and this is generally thought to reflect its having switched to low-cost devices instead of focusing on innovations consumers would want and willingly pay more for. Samsung emerged, struggled, survived, and thrived; and so on. But today the speed of innovation, of success and failure, is vastly higher than a hundred years ago.

The first smartphones able to connect to the web appeared in the 1990s and were clunky. BlackBerry and the iPhone, between them, changed the world. The smartphone has become one of the most important and essential technologies of both the developed and the developing world—with an estimated 2.5 billion smartphone users globally in 2019. In a number of African countries, the smartphone has allowed economies to leapfrog the technological progression that used to take developed countries decades. In Kenya, for example, much routine daily banking is done through the smartphone. Investment into landline-based phone systems has not been necessary. Smartphone-based health care is increasingly important. It can be argued that the smartphone represents the practical pinnacle of human ingenuity to date.

People's expectations have also changed. No longer do we accept a phone that is just a phone. We want one that is lightweight, has a battery that lasts forever, has unlimited memory, can monitor our health as well as our finances, can connect to the internet rapidly anywhere, act as a GPS system, survive being dropped into the toilet, unlock our car, manage our kitchen from a distance, turn off our lights, monitor our alarms, find itself or another phone when lost, be absolutely secure . . . Technology seduces us. Rather than being happy with what we have we want more. Fashion dictates that we need a new phone even when we don't. Industry wants us to have more. More capable, and often more expensive, models appear every year—all launched with fanfare and pizazz.

But these phones soon also performed other functions—they fed users' data back to the supplier. It was a Faustian deal that many other companies joined in on. These companies did not need to actually make physical things in order to succeed. Amazon, Google, and Facebook had a very different way of making money. And that all depended on the internet.

Wiring the World

Silicon Valley is named after the silicon chip, and is home to a high density of the companies of the digital age. The Silicon Valley culture is unique: the mega-barons of the Valley are using their wealth, power, and influence to shape the world to fit their vision—a vision of a world in which digital technologies rule supreme. The reshaping is promoted with much utopian trumpeting: that the digital revolution will bring better lives through a reduction in work and an increase in knowledge; that it will lead to longer, healthier lives; that the division of the world into nations will become obsolete. This techno-utopian view is reflected in many books, perhaps the most famous being *The Singularity Is Near* by Ray Kurzweil.[6] Kurzweil popularized the idea that technology will free humankind—technological singularity being the point at which machines can outperform humans and, by implication, develop their own new technologies without human assistance. A related concept is that of transhumanism, which aspires to technologies that will outstrip the biological constraints of our minds and bodies and could, some believe, lead to that age-old dream of immortality.[7] But these very narrow and technocratic views of our future are clearly unrealistic when considered more broadly. Such visions also obscure the real motive of at least some of the mega-barons—which is to accumulate more wealth and more power and develop a world that fits their vision.[8] We urgently need a dose of techno-realism. Even in high-income countries, let alone in low-income countries, a large percentage of the population is unable to access or gain value from these technologies—the global digital divide is still wide and could easily grow wider. We need to focus first on what is immediately ahead of us and on the impacts of technologies on ourselves without being caught in a futuristic speculative frenzy.

The internet originally emerged as a way to network research laboratories. Until the 1980s, most computers had hardwired terminals or were connected to terminals elsewhere in the same building by a dedicated telephone modem. The first more-connected system was the ARPANET, funded by the US Defense Department. Its first transmission of data, in what was called packets (referring to how data can be split and sent along a wire or fiber) from one computer lab to another, occurred in 1969. Vince Cerf and Robert Kahn developed a protocol that allowed transmissions to override the idiosyncrasies of different computer systems and enabled them to communicate relatively seamlessly with each other. This protocol became the basis for the modern internet, which emerged from the ARPANET as other laboratories were allowed to join in during the late 1980s. Cerf was to become the entrepreneur in residence for Google. Email was first used by academics; in the early 1990s it was opened up for commercial use. The pace of communication then accelerated dramatically.

In parallel with developments around ARPANET were those at a major pan-European research center in Geneva dedicated to basic particle physics—CERN—now home to the Large Hadron Collider. It too used the networking protocols that had been developed by Cerf and others. Tim Berners-Lee, a British professor who worked at CERN, recognized that the internet could also be used as a place where documents and information could be stored, retrieved, and linked electronically by hyperlinks. This was the start of the World Wide Web. In 1990 primitive browsers started to emerge. Early versions were soon replaced with more elegant creations—Safari, Internet Explorer, Firefox—accompanied by sophisticated, algorithm-based search engines such as Google.[9] The web itself continued to evolve. "Web 2.0" developments at the start of this century allowed users to do more than merely upload and download stored information—now they could work together remotely to generate content together in real time. This was the start of social networking, blogs, and wiki sites. The internet had radically changed how humans communicate, and we needed to adapt to this new, evolving environment and the group structures and digital identities it is continually creating and re-creating.

In his book *Grooming, Gossip, and the Evolution of Language*, the evolutionary biologist Robin Dunbar suggested that gossip has long been important to our evolution of group identity, maintaining the group's structure and creating a way to manage and control social norms and effect the social exclusion needed to deal with freeloaders.[10] He argued that in some ways gossip was the human counterpart of grooming, which plays a critical role in sustaining group structure and cohesion in many primates. When email arrived, with its capacity to interact with multiple people simultaneously, electronic gossip was sure to follow. Organizations found that their employees generally no longer confronted interpersonal issues directly by walking down the corridor to talk to colleagues—instead they preferred to send an email. It also soon became clear that emails could magnify issues out of proportion—social media was in effect being born. Chat rooms and blogging sites had appeared by the late 1990s, and by the mid-2000s there was a raft of social media platforms—YouTube, Facebook, Twitter, and an increasing number of more focused systems such as LinkedIn and Tumblr. Our evolved need for social connectedness and gossip morphed into communication and even courting via cell phones and their apps. Some forms of gossip soon morphed into the increasingly anonymous and destructive cyber-bullying and *ad hominem* attacks that dominate on social media today, especially on platforms such as Twitter. Reputations, real or otherwise, can be destroyed in a tweet.

The development of social media has led to dramatic changes in concepts of privacy that had been traditional (at least in developed countries). What we once considered private is now shared by many people on Instagram or Facebook. Sexting is now common. Anonymous *ad hominem* attacks dominate some forms of social media. Digitalization did not create these issues—gossip and its prurient dimensions long predate the internet—but it has given them a more central position in the lives of millions of people.

The digital transformation of our communication and access to information is the result of a clear symbiosis: our own desire for convenience and rapid access to information and gossip, and the interests of the platform companies, which can make huge sums of money out of our desires. The power, influence, and resources of platform providers have grown astronomically, to pervade every corner of society and, increasingly, of political discourse. Apple has built its dominance on apps and iTunes, Google on its search engine (which conducts real-time virtual auctions to place advertisements in your search results), Amazon on electronic selling, and Facebook on hosting social media. And their power depends very much on their scale. The larger they are, the greater their connectivity and data, the greater their power but also the greater their utility to the consumer.

But how are they really making money? By mining the data about us as users. Amazon exploits its user data by directing recommendations to the consumer using artificial intelligence (AI)-based algorithms that align our buying habits with that of others who purchased a similar range of products online. Apple does the same with iTunes and through services to the providers of the hundreds of thousands of apps that they sold or hosted. Search engines enable tailored advertisements to appear in a user's search results, for which advertisers paid the search engine company a fee for each click on the ad. These platform companies became financial gold mines. And Facebook, Google, and others have become powerful in ways well beyond their earnings alone.

These platform companies capture our data. They use it for enormous commercial advantage, directly and indirectly through AI-directed advertising sales and partnerships with other providers of services and products. And they sell information or access when a good opportunity arises, as, most notoriously, in the case of Cambridge Analytica.[11] In this scandal, a Cambridge University researcher built a quiz for Facebook users that enabled him to harvest their data and the data of all their Facebook friends. The data from up to 87 million Facebook users were exposed and sold on, including to organizations that may have influenced the outcome of the UK vote on Brexit and the 2016 US presidential election.[12]

Data of all types have become the most powerful currency and asset and are increasingly managed beyond jurisdictional control. Who owns these data—the person to whom the data relates or the platform company who gathered them? Most of the data are derived from the individual using the internet in good faith. We click "yes" to some legalese on an application—we generally have no choice, if we want to use a service—and we thereby give our details to a data bank. What rights do we have over our data? What are a government's rights to data about its citizens, and how should they regulate access to data by their staff and others? What are the boundaries of use and misuse, and what does privacy mean in the data age? These are all important questions that have gone largely unanswered.[13] One company in China claims to have more than two billion identified faces in its database and is using that data to build their facial recognition software. How did they get access to so many faces?

Recently Facebook published, in one of the world's most prestigious scientific journals, the *Proceedings of the National Academy of Sciences* (USA), a highly contentious piece of research the company had conducted. It was accompanied by an editorial that expressed the dilemma of the journal over whether to publish the paper, given the very dubious ethics behind it. Facebook had, without any external oversight, manipulated the data provided to over 100,000 of its users so that they saw either more positive or more negative information. The mood effect produced as a result was expressed in the blogs of those who received the manipulated messages.[14] So Facebook has demonstrated that it could surreptitiously manipulate the attitudes and feelings of individuals, and then their actions, on a grand scale. Facebook apparently had little, if any, interest in the implications of their actions. This could be seen as the digital equivalent of the infamous study carried out on African American sharecroppers in Tuskegee, Alabama, from 1932 to 1972 by the US Public Health Service, in which blood samples were taken from men with or without syphilis in order to monitor the progress of the disease. The study was conducted without informed consent, and without explaining to those infected with the disease that they would never receive treatment for it.

Should the responsibility for setting ethical standards and ensuring that they are upheld be left in the hands of an unregulated mega-company such as Facebook? If not, how might they be regulated? We are in new territory. The power such technologies give to, for example, the owners of Facebook, who have access to data on over 2.5 billion people, and the potential to manipulate not just their purchases but their attitudes, thoughts, and motivation through algorithms, gives them the opportunity to control people on a scale far, far greater than ever available thus far to the most autocratic dictator. Furthermore, it is now clear that such platforms can manipulate political decision-making, as well as economic forces, in ways that violate what we have traditionally understood to be our autonomy and conscious decision-making—our free will.

The manifest advantages of this digital transformation were clear from the outset. But has everything about it been to our advantage?

The digital world we have created through our ingenuity is now second nature to our species. It is an integral part of the niche we have created and continually modify. The modifications are remote and clearly do not benefit us all, and they certainly are very far from creating a stable society. The ingenuity which allowed us to become such a successful species in survival terms now seems to be threatening us.

The digital transformation our societies are now undergoing is still in its early years. Klaus Schwab, the founder of the World Economic Forum, calls it the "fourth industrial revolution"—the first being the industrial developments of the late 1700s and the 1800s, the second being the development of assembly-line manufacturing at the beginning of the 1900s, the third being the computer revolution of the second half of the 1900s.[15] And each of these transformations has resulted from our ingenuity and cultural evolution, and their sequence reflects our drive to continuously modify our lives. Braden Allenby, a scholar of technology and ethics, has said, "Any technology of enough significance to be interesting will inevitably destabilize existing institutions, power relationships, social structures, reigning economic and technological systems, and cultural assumptions."[16] The issue is whether this current, digital transformation is substantially different in its implications because of its speed, pervasiveness, and breadth of applications and consequences.

We think the digital transformation will be even more impactful than many expect it to be, and that it will require deep and broad reflection instead of the largely passive responses that are now common. Much of the discussion of this transformation has focused on the future of work and the impact of digitalization on productivity.[17] But the digital transformation is affecting far more. It is pervading every aspect of individual, family, and group life.[18] It is arguably the most important change in our evolution and in human society since the development of agriculture and living in settlements.

Digital technologies are expanding their impact in a number of ways that continue to evolve. First, they enhance access to data and information via the web and search engines. Second, communication is faster and more targeted, whether it is financial, commercial, social, or an interaction between an individual and the State via email and the internet. Third, connectivity now has an additional meaning. As the "internet of things" evolves, everything is becoming connected to every other thing, in the name of convenience and utility. Similarly, digitalization and microcomputing lead to the potential for all sorts of monitoring, whether through a smartwatch, a health tracker, or automated analysis of CCTV coverage. Such practices raise a multitude of issues: Who is doing the monitoring? Who is being monitored? And for what purpose? Is this monitoring enhancing or diminishing our security? Certainly our privacy is being further compromised.

Closely linked to all these developments is the emergence of AI and machine learning. Computational technology, both in hardware and in software, has developed to the point where microchips can perform calculations at such a speed, and in parallel, that real-time analyses of options can be made. The programs can look back at prior data, such as previous events or scenarios, and use that both to make decisions and then to refine those decisions. AI is all around us—it is why Google seems to know what advertisements to put in front of us, why some news stories seem to pop up at the top of our Facebook feeds, and why Book Depository can tell us which books people like us want to buy. AI is increasingly used in areas such as a decision on whether to give us insurance, or determining where police officers on the beat would be most effectively deployed.[19]

With AI comes the potential for much more advanced robotics and autonomous machines. It is claimed that these technologies will lead to large increases in productivity and that this will change the nature of many traditional jobs. The "future of work" is now a core policy issue for many governments. It has many dimensions. Will automation, robotics, and AI put people out of work? For those who have been displaced, will it be possible to retrain them for new forms of work? Enthusiasts suggest that the issue of unemployment is overstated, because these jobs will be replaced by other kinds of new jobs, and so on. But can retraining cope with those who are displaced? Even if workers can be retrained, will the new jobs suit them? How should people be educated for a world where change will be continuous and the concept of a skill or job for life will be meaningless? How will people be able to plan their lives? What will be the spillover effect on our mental health or our life-course planning? How will tax bases need to be designed—should robots pay income tax? Will the gig economy become the norm? Will current forms of social welfare need to change? We are already seeing early experiments on concepts of a universal basic income—all emerging from the policy dilemmas arising from consideration of the future of work.[20]

In a similar vein, what will be the effects of autonomous vehicles? What will the future hold for truck and taxi drivers? Or for ordinary drivers, who will be ceding control of their vehicles to software? And there are ethical and legal issues—such as, Who is responsible when an autonomous vehicle has an accident? All of this is new ground, and the technology is outstripping ethical and policy considerations. The issues are even more serious when we think about the development of fully autonomous weapons. Currently a number of nations conduct warfare using drones. But to date these have been under human control. Some countries are developing fully autonomous weapon systems that might be enabled to make independent AI-based decisions on when to attack and who to kill. And they will do so quickly, because the opposition's AI systems will work just as fast, so a human override may not be possible. The ethical questions have enormous implications. International treaty discussions have started in Geneva but are not progressing well. Such treaties are not perfect in preventing illegal actions, as we have seen for chemical weapons, but they can be inhibitory.[21]

We are only at the beginning of our online revolution, and the debate over its potential and the speed of its impact has hardly started. But already we can see different camps being formed. On one side there are the techno-enthusiasts of Silicon Valley who believe the human condition will be greatly improved by the approaching "singularity," the hypothetical time in the near future when the capacities of AI will lead to runaway technological evolution, artificial computer-embedded superintelligence will outstrip the need for human engagement, software will upgrade itself, and our relations with machines will be changed forever. Yet others who have been termed alarmists have argued that our ingenuity might lead us to create an AI which no longer needs us and, if so, what does the future hold for the evolution of human intelligence and indeed for our species? Even though there might never occur such a technological singularity in the form of a superintelligence independent of human understanding and control, and most AI experts see this as eons away, the issues raised by the concept are real and have inspired increasing concern. There are those who believe that much greater consideration needs to be given to the relatively unregulated nature of these new technologies, of this aspect of our ingenuity, and the ethical issues they raise. For example, Lord Martin Rees, former president of the United Kingdom's Royal Society and one of the world's great cosmologists, envisions the potential of AI in a very negative and frightening way.[22]

What will the future hold? It is impossible to know, but a key feature of this transformation is that innovation is proceeding quickly and is driven by multiple sources, often emerging directly from the private sector. Given that the uptake is also very rapid, it usually happens without societal or policy oversight. We see a growing gulf between the technologies we have created and social oversight on how those technologies are used. Technologies being developed will advance the use of AI far beyond its current uses, and augmented and virtual reality will take things further. With the power of these technologies to present the false as true, the nonexistent as real, how will we be able to tell what is real or not—for instance, when highly convincing media coverage claims to reveal that the leader of a country is making a threat in a conflict situation? Serious discussions are ongoing about the impacts of such "deepfake" scenarios.[23] And we must also consider the changing relationship between technology and ourselves. We are just at the beginning of this digital revolution, but it is already clear that it is changing our concepts of reality, something that is fundamental to understanding what makes us who we are. The potential for body implants of various sorts that marry biology and electronics to monitor our physiology, treat disease, or augment or otherwise alter our capacities, will likely go well beyond Google Glass, heart monitors, or wearable insulin pumps. Various forms of brain implants—technologies to alter our brains—are no longer merely the material of science fiction.

The pace of change is fast, and the potential impacts of these new technologies will likely be extraordinary. It seems beyond the capacity of any single government or society to address. After all, the world is still coming to grips with the consequences of the global financial crisis of 2008, which itself was in no small part a reflection of the digitalization of the global financial system.[24] Clever people with clever computers connected via the internet could change the financial system in ways that policymakers or even the financial community would not be able to understand, control, or regulate, given the complexities and interdependencies they've created. In 2008 the interconnectedness of the global financial system was such that, when one relatively small part of it collapsed in the United States, the ramifications could not be foreseen, predicted, or comprehended, let alone controlled. Compounded by computer-based decision making, which is now the norm in trading in the financial sector, decisions were made that have had persistent global consequences. The financial implications, especially for already disadvantaged sections of societies, have fueled angry political movements in many countries. The global economic policy community has proven to be very limited in its ability to move beyond the global financial crisis and to create an environment that is reassuring, and the social and political consequences are obvious. Self-interest can inhibit progress, as we have seen with climate change. The implications for social cohesion, for the way many of us live our lives, and for national and global politics have been pervasive and sustained. The consequences have included the relative collapse of the middle class in many liberal democracies, which in turn has played a large part in the rising anger at growing inequalities and the fueling of populist politics.

Digital Realities

While almost everyone would agree that the digital transformation has been a tremendous boon to communication and access to information, the internet does not allow us to discriminate between reliable and unreliable information. Anyone can place information on the internet. Its reliability depends entirely on who put it there and why. Even Wikipedia, which was established with the intent to be a bottom-up, neutral, and reliable information source, has much information of doubtful accuracy and validity. There have now been a number of studies of the reliability of medical information on the internet—important because many people use it as their primary source of information about their health—and it is concerning how many sites of apparent validity have erroneous advice. Sometimes this is not from bad intentions but because of the lack of expert input and independent review of the content.[25]

Political organizations and individuals or groups with strong predetermined positions and with the intent to enlist support or disrupt society have become experts at intentionally posting false claims on the web that have all the appearances of high validity, either to advance their cause or to manipulate opinion. The ease of posting targeted messaging through social media encourages this approach. It has been a common feature of the climate change denial movement and has recently risen to the level of an art form in the political arena, where it has been used to influence voters and potentially to undermine national governments.

Manipulative and targeted messaging reached a zenith in 2017 with the slogans "alternative facts" and "fake news" (the accusation applied to anyone who produces evidence, no matter how robust, which does not support a particular view).[26] And, perhaps most tellingly, "post-truth."[27] The internet has made available an almost unlimited amount of not just genuine information but also false claims, and this has been accompanied by a critical and dangerous shift in attitude—many people see no need for the synthesis of evidence and evaluation of derived conclusions, which are the roles of experts. Respect for wisdom is being replaced by a hubris that assumes access to information is sufficient, even though that "information" might be false or biased to fit a pre-established point of view. This is the "post-expert" world that was highlighted in the United Kingdom's Brexit referendum and in many politicians' dismissiveness toward climate-change science.[28]

None of these phenomena are unique to today's digital world. Political propaganda, disinformation, and manipulation of facts have been widespread ever since societies first developed hierarchical power structures. But there is no doubt that the internet and social media have made such manipulation of knowledge and opinion easier and more pervasive. Legislators have done little to oppose these trends. Is it even possible for any country to devise effective solutions to these problems? Is it too late? Has the reach of digital media become too wide, now being global and not confined within any country's boundaries?

There are other looming concerns: cybersecurity and cyberwarfare are now central to national strategies. Corporations and individuals worry about hacking, phishing, loss of privacy, and disruption. Company after company faces security breaches and reputational damage as customer information is stolen to be sold on the dark web. National secrets turn up in the media, institutions are held to ransom by ransomware; digital viruses and worms invade our computers. Digital hygiene—the care and maintenance of one's online accounts, information, and reputation—has become as important as personal hygiene. The emergent "internet of things" will make the danger of hacking, malicious manipulation, and e-ransoms much more personal to everyone who becomes that connected. And it also has the potential to drastically change the relationship between citizen and state as is evident in the emergence of social credit ratings in China.

Social media, by definition, allows us to select the views we want to hear. It is inevitable that the selection of friends and followers is linked directly to a matching of worldviews. We all like to share ideas with people who are likely to agree with our ideas and viewpoints. "Group-think" was an important source of social cohesion in human evolution. Thinking differently—whether it was rejecting authority or the mores of group ritual and identity—would lead to exclusion.

But whereas group-think in the age of simpler social constructs, such as those reflected in Dunbar's analysis, included a broad range of views across the group, now it is different. We want to be liked and to be part of a social group. We tend to mold our views to those of our networks. But now we can select our networks electronically—our allegiance is shifting away from physically close groups, where inclusion requires compromising with people with diverse views, toward "virtual" associations of people who hold essentially identical views, irrespective of where they are. This is the "echo chamber" of social media.

This echo chamber is not unique to social media. Digitalization and other technological developments have produced a major shift in mainstream media. Many of us read newspapers or news summaries online and watch streamed content such as films, concerts, and sporting events, and select these according to our preferences, rather than watching traditional television stations, which give broader coverage. But nowadays even traditional TV channels feature content aimed at viewers with a particular point of view. Those of a certain political persuasion will see Fox News, or CNN, or Al Jazeera as an accurate reflection of the world, whereas others will see the same channel as biased. The same would be true of almost any mainstream media outlet—print, radio, or visual—that adopts a particular framing to appeal to the views of a specific audience or readership. If customers' views are endorsed, they are more likely to continue to access or purchase the resource, and of course this is of great value to the news outlet and advertisers.

A number of experiments in the behavioral sciences have shown that values and preexisting biases, rather than facts, determine largely what we believe. Even if information is provided in the format of scientific data, we tend to interpret it in different ways, often confirming our preconceived ideas. Sometimes, offering more data only reinforces our views rather than challenging them. For example, when the same information on climate change was presented to deniers or environmentalists, both saw the same data and the same presentation as supporting their views and reinforcing their position.[29] Another example is the modern tragedy of the "anti-vax" movement, which seized on flawed science that linked vaccination against common childhood infectious diseases to various conditions including childhood autism.[30] This misinformation has been promulgated by many entertainment celebrities (we now live a world where celebrities seem to be viewed as having more authority than actual experts). Despite the total refutation of any association between the measles vaccine and autism, the manifestly false anti-vax claims are continuously refreshed via social media and have led some sections of the community to not allow their children to be vaccinated. In many countries there is a resulting resurgence of measles, a disease that can harm the unborn baby and be life-threatening in children. The success of immunization depends on protecting the vulnerable child but also on the concept of "herd immunity"—that is, if a large enough percentage of the population are immunized, the virus cannot persist in the population and anyone not immunized thus has a very low risk of being infected. There is an interesting analogy here with the issues of within-group cohesion and the threat posed by freeloaders. For vaccination, the freeloader's position is to not allow their child to be immunized, in order to avoid the very small risk of side effects of the vaccine. Provided that almost every other child is immunized, the freeloader's child will benefit. The strategy fails, however, when there are too many freeloaders, because then herd immunity is diminished and the risk of a measles epidemic increases, which will affect many parents and children in the population. Group cohesion has then been shattered.

Digital Politics

This polarizing effect becomes increasingly obvious when we turn to issues of social cohesion and political ideology. Inherently people have differing worldviews regarding what is fair or egalitarian and how society should operate. In the past these views most often were acquired from the family's social and economic context, similar to the way religious beliefs are acquired. But social media now encourages us to seek peers who share similar worldviews, and so these views become reinforced rather than challenged. This may play a role in the greater polarization of political views, from far left to right, which many countries are facing. When social discourse was primarily face-to-face and less self-selected, the potential for hearing views contrary to our own was greater.

Such polarization undermines the efficacy of the democracy on which our societies depend. Winston Churchill is often said to have remarked that "democracy is the worst form of government, except for all the other forms."[31] Most of the developed world has adopted some form of democracy based on elected representatives of political parties—whose ideologies are generally shaped around their views of economic, environmental, and social priorities for their country. For much of the last sixty years the political trend was centrist—seeking pragmatic solutions that can withstand political cycles and that generally try to accommodate the different worldviews of their voting communities. But the digital transformation has both directly and indirectly affected this consensual approach. The directly disruptive effects include polarizing the nature of the relationship between citizen and representative, and the ability for worldviews to be manipulated. The indirect effects include a shift to the use of rapid and often selective opinion polls by politicians before decisions are made, which drives a short-term-response form of politics.

The techniques of AI, first applied to help us purchase what we are looking for on websites such as Amazon, are now being extensively used in the political domain, including the increasing use of targeted messaging via Facebook and other social media. What does "democracy" mean when our minds are subtly manipulated via social media by the very individuals desiring power? Or when political decisions are made on the basis of instant polls rather than through the intermediary of representatives debating issues in depth before arriving at a majority decision? Some see the digital trends as being closer to the idea of participatory politics and argue that direct participatory democracy may be more effective than representative democracy. But isn't the digital realm even more vulnerable to manipulation? Others would argue that we need to find a way back to the ideals of representative democracy in which elected representatives are expected to rise above the fray and look beyond their immediate interests toward the broader interests of those who elected them.

The Government's Role

Governments are challenged by these developments. They have difficulty protecting privacy and data, and their protection of individual citizens has been compromised. Laws of libel and slander have been greatly weakened by the internet and in particular by the empowerment of anonymity. What is the responsibility of the platform company? Does it simply host the site or is it a publisher? If the latter, does it have the responsibility to edit and restrict content on the site? Whereas governments have had some role in regulating privacy and interpersonal respect through such things as libel and slander laws and regulation of pornography, the globalized nature of the internet greatly diminishes any government's power. We face serious political mismatch. As social animals, we have a fundamental need for a regulatory framework that gives order to our lives and gives us a sense of position in society. But the digital world now determines many aspects of our lives, and we no longer have any control over this or even any real ability to influence it.

Our fiscal and monetary systems are at risk, too. Globalized digitalization has enabled instant financial transfers to avoid paying taxes on wealth. It has been difficult to implement global arrangements to counter this, largely because tax-haven countries don't want to counter something they benefit from. There also is increasing concern that the development of the gig economy (which has become possible only because of AI and the internet) might undermine traditional taxation bases—something that might be accelerated by the growth of automation and robotics, if a pessimistic view is taken of the future of work. And digital currencies, which depend entirely on the opportunities created by blockchains, could potentially undermine things further. Digital currencies such as Bitcoin create a way to hide transactions—much of the dark web's trade in drugs, weapons, sex, and people-trafficking is possible because of these cryptocurrencies. If any of these electronic currencies were to gain a dominant position, it would create the means for citizens or companies to avoid financial regulations and even lead to countries losing monetary and fiscal control, undermining key tools of government. If the type of unbreakable encryption that quantum computing might possibly bring is combined with cryptocurrency, this could become very threatening to national governments.

But governments themselves are using digitalization to increase their powers. Like companies, governments are interested in capturing data and using it. Immigration and customs authorities rely heavily on the internet for surveillance. So do police and security services. Different countries have taken very different approaches. Some, like North Korea and China, use high levels of state surveillance to control what their citizens see and how they can act. But whether in a liberal democracy or in an autocratic state, the issues of use and abuse of data by the government are real and need ethical oversight. There are proper uses of data by governments, such as to provide national security, public safety, and services for its citizens. But there needs to be transparency about what data are being collected, how they are being used, and what safeguards are in operation. This is new and fast-moving territory for policymakers, and yet it must be a priority if the digital transformation is to be used to societal advantage and in a way that creates more societal cohesion and equity rather than less.

Data can indeed be an enormous help to governments in providing more efficient and effective services. It took us less than twenty-four hours to get an electronic visa from New Zealand to visit the gorillas in Rwanda. In countries such as Estonia, virtually all interactions between citizen and state are now electronic—this is the realm of e-government. Data can also help governments explore complex environmental and resource issues and find options for policy development. Without digital technologies, countries would be far less equipped to deal with issues such as water quality, climate change, biodiversity, and population movements associated with infectious disease.

Traditionally, areas such as health, education, justice, social housing, welfare support, pensions, and employment have all been separate domains of government, and the links between them have been treated rather superficially instead of in a genuinely structural way. Even though digitalization involves important issues in ethics and governance that need continual review, the potential to use linked data to understand how people passage through life and how to best assist them is enormous. A number of countries are already attempting to do this. For example, New Zealand has developed the

Integrated Data Infrastructure—a systematic approach to linking administrative data across government agencies—which has the primary goal of understanding complex interactions that affect people in very different ways.[32] It aims to inform policymaking in areas like intergenerational poverty and child welfare, and recognizes that often what is seen as purely an issue of health, welfare, or justice has its origin in much more complex societal considerations. These need to be addressed if the government is to do better by its citizens. But even here there are many ethical and societal concerns that will require considerable and ongoing effort to be addressed.[33] This is the confusing potential of big data and AI.

But data and sterile analysis are not enough. Just as we have raised concerns about the use of the internet as a source of unfiltered data and claims that lead to a rejection of the need for experts to interpret them, so it is with "big data" analytics. Properly used, big data will be a great tool for benign government; poorly used, they will lead to misinterpretation and bad decision making.

Perhaps paradoxically, surveys have shown that we prefer companies, rather than governments, to have our data.[34] This may be partly because it is increasingly difficult in many societies to function without use of the internet for shopping and services. The implied contract is that the company is doing something for us and in return we have no choice but to share our data. Usually we are aware that our data is being collected, used, or sold on by companies, but we accept that deal because of the immediacy of the services provided. The European Commission has recently taken some early steps toward protecting data with their General Data Protection Regulations.[35] But this is probably just one finger in a dyke that has many leaks, and in any case most citizens are not comfortable with governments having control over data. Some fear that an Orwellian world of state surveillance is upon us. Will the government use personal data to help us, or to control and manage us? Most of us would approve of its use to catch rapists and murderers—as when California police recently succeeded in using genealogical databases and genetic information to track down a serial killer through genealogical data of his relatives even though his own genetic data were not in the database.[36] But what about the government using databases to detect a welfare fraudster or people overstaying their visas? The answers get more complicated and value-laden. Do we want "Big Brother" watching our private lives and then using it to reward or punish us? Indeed the social credit score system in China appears to be an example of such an approach already being developed. Our cell phones track where we have been and when. Digitalization is changing the implied contract between citizen and state, and it is urgent that governments establish a clear and transparent relationship with their citizens over data use and governance.

The digital transformation is profoundly changing the human condition in many ways—from how we live our individual lives, to how we relate to each other and how we interact with our governments. It is changing the power structures between the private sector and the state and has implications for the nature of capitalism itself.[37] It is a powerful example of how our technological evolution is changing our culture and thus indirectly our biology, for it produces a dramatic change in the niche which we have modified. It is now clear that it is damaging our mental health and probably changing the way children's brains develop.[38] It is changing the way we engage with each other. With all this technologically created culture, are we winning? In Chapter 8 we will look at the balance sheet.

8 | COST

THROUGHOUT THIS BOOK we have discussed how we evolved with unique ingenuity, devising and implementing new technologies cumulatively and at an accelerating pace. This has made us what we are. Our ingenuity has enabled us to populate a wide range of environments. Our innovations have assisted with our evolution from hunter-gatherers, to pastoralists, to settled agriculturalists, to city dwellers. They have changed not only our physical world but also the world we inhabit virtually. Unlike any other species, we have an evolved ingenuity that allows us to be continual and cumulative niche modifiers. So far it has been a highly successful strategy.

But any strategy to change nature comes with a cost, and adopting one strategy often precludes using some others. The peacock evolved with his spectacular tail to attract the peahen, and as a result will never be able to be a migratory bird that could escape a substantial change in his environment such as an increase in predator numbers or climate change. The termites in Litchfield National Park in Northern Australia cannot move far from the safety of their compass mound, and nor can they take it with them. The downside of a constraining strategy is most likely to become apparent if the environment changes drastically, or if a strategy developed in one environment has to be used in another. If the strategy is embedded in their evolved biology, then in that new environment the individuals of the species will be "mismatched" to that environment. Then the species must move, adapt, or die.

In our previous book, *Mismatch: Why Our World No Longer Fits Our Bodies*, we argued there are two types of mismatch, which overlap and can coexist.[1] The first is evolutionary mismatch. This occurs when the environment changes either too much or too fast for biological—that is genomic—evolution to keep up. In this chapter, and in Chapter 9, we will be thinking about the implications of such mismatch for our own futures—which, given our continued innate drive to innovate, may get worse.

In Chapter 3 we also discussed the second type of mismatch, developmental mismatch, because during development our bodies make predictions about the world in which we will live, hopefully to survive and reproduce in it. The predictions are based on cues we receive during early development from our environment, from our parents, and more widely, and they change how we develop both in terms of our anatomy and physiology. These predictive adaptive processes are likely to be why, like other organisms, we have retained the mechanism of developmental plasticity. Many of the underlying mechanisms are epigenetic—mechanisms we discussed earlier. But sometimes the early life signals can be wrong, either because the environment in which we develop is not representative of our later environment, or because the environment has changed rapidly over a generation.

Since publication of our book *Mismatch,* we have continued to ponder these two forms of mismatch, which are not entirely independent of each other. Much of our research and our work in health policy and education had been focused on the question of the chronic, noncommunicable diseases (NCDs), which now account for more than 70 percent of deaths across the world every year. The biggest killers are cardiovascular disease, cancer, lung disease, and diabetes.

One of the major risk factors for all of these NCDs is obesity. In the campaigns to reduce levels of obesity over the last few decades, most attention has been paid to two underlying "drivers" of the problem: genetics and lifestyle. The fact that the problem of obesity has now reached epidemic proportions, and affects just about every population in the world, shows that we, as a species, have not yet found the solution. The obesity epidemic is a dramatic example of our inability to cope with our present environment—one we have created for ourselves.[2] We moved away from self-sufficiency in food supply to traded food, shifting, particularly after World War II, to a more industrialized production of energy-dense foods, rich in high-fructose corn syrup or added sugars, unhealthy fats, and salt, and low in fiber. Our energy intake is now out of balance with the energy expenditure of urban living. It is exacerbated by the fact that the most energy-dense foods tend to be cheaper and more convenient—explaining why those who have less income are more at risk of obesity. This is not only an issue of so-called developed nations, because in India and sub-Saharan Africa the rates of obesity, especially for the urban poor, have been rising rapidly. Other aspects of our lives, especially Westernized urban lives, contribute to this mismatch. City life and industrialized food production generated the fast-food industry, which is dominated by energy-dense, highly processed, and nutritionally unbalanced foods that are tasty and cheap and often sold in large portion sizes. The time saved by consuming these foods gives an added incentive to purchasing them. The cost of choosing to save time and have easy access to fast food is a high-calorie, unhealthy, processed meal. The levels of physical activity required in the early eighteenth century at the start of the Industrial Revolution were much higher than they are today. From the manual work required in farming or horticulture, in shop or factory work, to domestic chores and transport, physical movement was an integral part of human life. Many of the rural and urban poor probably had low levels of nutrition and calories, and it is likely that their diets would not supply enough energy for more than a few hours of hard labor each day. It is quite different today.

Another very worrying example of evolutionary mismatch is gestational diabetes, which occurs in pregnancy. Its prevalence is rising rapidly in many countries, especially in Asia. It is associated with obesity in the mother, and begins quite early in pregnancy. It conveys greater risk of obesity and diabetes in the woman after giving birth. Importantly, it is also associated with excessive accumulation of fat in her developing fetus, making birth difficult and even dangerous, and with potential metabolic problems for the newborn infant. The newborn infant in this situation has more fat cells when born and is more likely to become overweight and develop diabetes later in life. This is because obesity is primarily a function of putting more fat into the fat cells, the equivalent of having an expanding gas tank in a car that is able to store increasingly larger amounts of fuel. We believe gestational diabetes was relatively rare until recently because of two evolutionary considerations: First, obesity in general is largely a recent phenomenon associated with the introduction of processed foods and urbanization. Second, the placenta, which is responsible for controlling the passage of almost all nutrients from the mother to her fetus, does not regulate the passage of glucose. In our evolutionary past, humans would rarely, if ever, have experienced the sudden spike in blood glucose associated with such practices as consuming a fast-food meal and a highly sweetened beverage. So there was no need to evolve a placental control mechanism to prevent exposure of the fetus to high levels of blood glucose. But now, with our high-energy, low-nutrient diets and no control mechanism to protect a fetus from high levels of glucose in the mother's blood, both mother and her fetus are at the sharp end of an evolutionary mismatch.[3]

Such thinking emphasizes how environmental effects in our early life—in this case, exposure to high levels of glucose from the mother's blood—can have dramatic effects on our development. Developmental plasticity produces a range of effects on many systems in a developing baby, not only in terms of its fat deposition. There are effects, during fetal life, on the developing heart, kidneys, and pancreas; on metabolic control; and on brain development, affecting emotion, cognitive functions, appetite, satiety, and stress responses in the infant. Practically every organ system is affected in some way. The changes in structure can be permanent; for example, the number of muscle cells in the heart and filtering units in the kidney is established before birth and cannot be changed later. In addition, this developmental plasticity operates even in response to the normal range of cues all fetuses receive in early development: in other words, these are not only effects that occur under extreme conditions such as maternal starvation or gestational diabetes.

Developmental plasticity sets our biology to function best in a "predicted" environment, an environment similar to that which shaped our earlier development.[4] There are many obvious examples of how our world has changed between generations—diets and physical activity being just two—making the prediction inaccurate. So it is easy to see how, in the face of such changes in environment, developmental and evolutionary mismatches can result in greater risk of chronic disease.

So What's New?

As researchers and educators, we are often called upon to give lectures or seminars on our work. These talks might be to our academic colleagues, some of whom are involved in research in related fields and some who are not; to students studying medicine, nursing, midwifery, or biological subjects; to officials from government and nongovernmental or philanthropic organizations concerned with health; and to more general public audiences and the media. We can usually tell if the presentation has been well received by the nature of the questions asked by members of the audience afterward. But recently we have noticed a pattern emerging in these questions. For example, we often give talks that explain the concepts of mismatch, using the examples of the epidemic of obesity and the global challenge of NCDs to illustrate the point. But quite often questioners have asked about other global problems and their impact—mental health, or climate change, or pollution, or the effects on children of high levels of screen time, or the impact of social media and so on. How, they might ask, does that fit in from the perspectives of evolutionary and global health? Or a questioner might ask: Won't we be always able to rely on inventing new technologies to cope with these mismatches?

Our answers are embedded in this book. Human ingenuity in developing and applying new, advantageous technologies can have negative effects—and the solution generally has been to develop a new technology to fight back. For example, widespread extraction and combustion of fossil fuels led to massive economic and social change from the mid-nineteenth century, but at the cost of global warming with its consequences. The commonly proposed and only very partial solution has been to devise new energy technologies, such as wind or solar, plus hoping for developments that will allow more effective carbon capture and storage to reduce greenhouse gases. The global response to climate change largely assumes that the problem can thus be solved based on imagined technologies that do not yet exist or on developing technologies that are not yet significantly operational. A cynic might be reminded of the comment that has been made about nuclear fusion for the last thirty years: it is always another thirty years away. Clearly, addressing climate change needs a much more holistic approach—one that encroaches on many vested interests whether at an individual, corporate, or national level.

Ingenious new ways of making energy-dense convenience foods fueled the obesity and diabetes epidemic—now myriad medicines and machines are needed to deal with the consequences. It was recently believed that bacterial infections had been defeated with antibiotics, until the rise of antimicrobial resistance and superbugs that require finding new ways to fight them. And lastly, the digital transformation brought email and access to immense stores of online information, but led to hacking, cyberbullying, loss of privacy, and many more dire consequences. Like for these examples, all technological innovations have downsides as well as upsides. In the past the downsides have been dealt with through ingenious creations of still more technological solutions. What appears to be different now is that the pace of new invention may be outstripping human capacity to respond and manage that balance of cost and benefit. It is time to take stock of the situation.

Between Ourselves

Humans are social animals and, like other social animals, we have evolved with behaviors and biological characteristics appropriate for the kinds of group structures that maximized our fitness. And as we have seen, the evidence suggests that our brains evolved for living in groups of perhaps no more than 100 to 150 people. There is a compelling argument that this combination of social structure and our ability to learn and to communicate via language allowed the knowledge and skills possessed by a group to be greater than that of its individual members.[5] Because ideas generate ideas, the innovative capacity of the whole group will also be even greater than the sum of its individuals' inventiveness. Innovation is heavily dependent on such collective knowledge, and, in turn, it enabled individuals within the group to have different roles, to perfect their skills, and to contribute to a more complex hierarchical social structure that still lies at the heart of our societies today. When the base of communal knowledge was expanded through writing, then printing, then the internet, the pace of innovation became greater still. We all rely on knowledge held within a group, rather than within our own brains, to navigate our way through our lives. Often we access this through several overlapping groups—at work, among friends, through institutions, and so forth. Today we are likely to primarily depend on the internet for communication and knowledge. Google, with all its covert filters and biases that select what information to give us, has become our almost constant companion.

The explosion of technological advances we have made as a species changed almost everything about our lives. Our innovations have changed how we move around our environment, how we communicate, and how we manage our health. They have changed our food production, the reproduction of our species, and even how we harm or kill each other individually or *en masse*. Whereas the Industrial Revolution primarily ended in the nineteenth century, the human technological revolution continues, and at a faster and faster pace. We seem driven to ride this accelerating revolution. Are we so dizzied by its speed that we do not see how it is changing our world as we rush along?

Cost and Benefit

In his book *Innovation and Its Enemies*, the renowned scholar of innovation Calestous Juma points out that there are both external and local influences that can explain the sometimes long delays associated with adopting many technological developments, and yet there are other developments we adopt almost instantaneously.[6] In general, he argues, people find change somewhat uncomfortable—and by definition new technologies herald change. At a practical level, we often see new technologies through the lens of cost and benefit. If we can see the direct benefit for ourselves, we are more likely to support a new technology. If, on the other hand, the direct benefit is not obvious or does not include us, then we are less likely to do so. We reflect these biases in our assessments of risk and what we see as the appropriate amount of precaution. All technologies involve some degree of uncertainty and impact beyond those perceived or claimed by their inventors. And concepts of risk and precaution are heavily influenced by contextual and cultural factors, which are often translated into policy and legislation.[7]

One technological development that many societies immediately addressed was the potential to make and grow genetically modified (GM) food crops. Different countries took different approaches. Some, like the United States, took a rather *laissez-faire* approach; others placed all sorts of controls on its use. Some countries, like New Zealand, effectively banned the use of GM processes outside the laboratory, for example in field trials. This is an unusual case where a process itself, rather than the application of that process, is regulated. And many of the regulations seem hypocritical. For example, in Europe GM crops are not grown in large amounts but they are imported as animal feed. The campaign against GM crops had many drivers. Some were purely commercial—some European companies were worried that US companies would get a step ahead of them and pressured politicians to be protectionist; some were philosophical—humans should not be 'playing god' by modifying other species; some were cultural—that GM will shift the balance between small and large farmers; and some were risk-based—concerns about health and ecological safety. The different views about these technologies originally arose from varied perceptions of who would benefit and who would bear the cost.[8] In the case of GM crops, the initial benefit was clearly for Monsanto, the company who first produced GM crop seeds. But the risk was seen to be entirely for the public. The situation now is very different, with the technology widely available even in low- and middle-income countries through both public sources and smaller companies. After three decades, the evidence for adverse health effects of GM is nonexistent and for environmental effects is minimal. Yet the main arguments still being advanced against GM crops remain related to safety rather than to any broader issues. The scientific consensus on the benefits of GM technologies is now far greater, yet in Europe, New Zealand, and some other countries, politics discourages further discourse on such technologies.

The next generation of genetic technologies is gene editing, which is a more precise way of modifying the genome and does not involve transferring genes from one species to another.[9] Again we are seeing great diversity in how the technology is perceived. Some countries view it as an intentional form of mutagenesis, similar to that occurring spontaneously in nature, and as therefore not needing regulation. Others view it as being no different from earlier forms of GM and regulate it in the same way. Yet there is a broad scientific consensus that these technologies could have a major role in promoting food security and in reducing the carbon intensity of agriculture. But different societies can take differing views on such technologies, and once those views are established, even if the evidence evolves, they are difficult to change.

Nuclear power is a very effective source of energy and is carbon neutral (apart from the cost of building the power plant and mining the fuel in the first place), and some see it as a logical step in addressing the challenges of climate change. Some countries use it extensively, some in a limited way, and others have rejected it completely. The issues that limit its use relate to safety, storage of radioactive waste, and security (because its by-products can be used for the manufacture of nuclear weapons). There are safety concerns that leaks could lead to an environmental disaster, such as at Three Mile Island or Sellafield; or to catastrophic accidents, such as at Chernobyl, where it was the result of human error, or at Fukushima, where it resulted from a perhaps predictable natural disaster. Newer reactor designs are considered much safer, but few countries continue to develop nuclear power, given the likelihood that public resistance to the risks it poses could lead to a political backlash. After Fukushima, Germany shut down its nuclear power stations, despite the relative lack of risk from natural disaster in that part of the world, and returned to coal power—despite the consequent massive increase in greenhouse gas emissions. Japan is building a series of "greener," coal-fired power stations. And coal-fired power stations are being built to meet global energy demands at an astounding rate despite the world heading toward unacceptable levels of warming.

In contrast, in the case of the internet and many of its associated technologies, the technology was seen as being of immediate and direct benefit globally, and broader issues of risk were little considered. Many countries rapidly adopted use of the internet and readily gave all sorts of corporations access to our data from which they make enormous profits. We have seen the technology greatly change the ways we think, communicate, interact, make friends, socialize, enjoy our leisure time, work, bank, do business, and interact with authorities. We readily accepted these conveniences of the internet, along with portable methods for accessing them—tablets, smart phones, and other devices. Beyond the telecommunications sector, governments did not interfere and there was no demand for oversight or regulation. But now we are starting to pay attention to other issues—such as privacy, data ownership, security, loss of jobs, monopoly behaviors of platform companies, and impacts on our democratic processes.

The perceptions of cost and benefit that drive decisions to reject or adopt a technology depend on a society's worldview, which can be very difficult to change. Science alone cannot change strongly held views, as we've seen with discussions of climate change, where both climate change skeptics and climate change believers can interpret the same sets of data as reinforcing their established views. This makes it hugely difficult to arrive at consensus or to address technologies in an adaptive manner, taking a risk-management approach. As we learn more about a technology, we should be able either to relax or impose more controls on it for societal benefit, but the combination of political inertia, vested interests, and our own fixed biases makes that difficult.

A further issue is that "risk" and "precaution" mean different things to different people.[10] This lies at the heart of the challenges we face with technological evolution. For scientists, risk can be expressed in actuarial terms. For the public, risk is generally perceived through the lenses of emotion and cognitive biases rather than in a purely logical manner. These two contrasting forms of thinking are sometimes referred to simplistically as system 1 (emotional) and system 2 (rational) thinking. In general our initial reaction to a situation is to make a rapid response using system 1. Then we might analyze it further using system 2. There are powerful evolutionary arguments for having these two forms of thinking.[11] An immediate response makes sense to avoid danger; the cognitive biases likely evolved to provide a simple heuristic for instant decision making. Analytical thinking is much harder work and requires a very different effort. So when we face a potential risk, we are far more likely to react to it using system 1 thinking. Suppose that 1 in 10,000 people exposed to a pest-control chemical might be at risk of a significant adverse health effect. If the chemical exposure level rises, perhaps 2 in 10,000 will be affected. The statistician will say that the health risk attributable to such chemical exposure remains very small, perhaps especially compared to health risks carried by the pests. But many people, and often the media, might construe this as a frightening doubling of the risk. Another example is the risk of flying. Even though flying is much safer than traveling by car, we are more likely to be scared of flying, and after an incident like a hijacking or an aircraft crash, fear and the perception of risk rise. In his book *Reckoning with Risk: Learning to Live with Uncertainty,* Gerd Gigerenzer points out that more people died from the increase in car accidents in the year after the World Trade Center attack on 9/11 than died in that horrible terrorist attack.[12] Air travel came to be perceived as very risky, and people shifted from commuter and short flights to driving long distances. More cars on the road for longer periods inevitably meant a greater road traffic accident rate. But politicians have a very different definition of risk. For them it is measured at the ballot box or, in autocratic countries, the chance of a coup or an uprising.

Whose Values?

We largely accepted the digital age without reflection. Therefore, unlike GM crops, it developed largely independent from regulation. But the nature of the internet also made it increasingly difficult to regulate. The anti-establishment culture of the internet in its early days, a culture still promoted by the platform companies as it suits their business model, kept it free from regulation. No one owns the basic infrastructure, which is distributed across all levels of jurisdictions with enormous redundancies, making unilateral regulation impossible. And this globalization enables users, such as financial and communication systems, to also avoid jurisdictional control. Unlike the transnational companies of the preindustrial age, such as the Dutch East India Company, or those of the industrial age, such as the powerful oil and automobile companies, these large global-platform companies have almost instant global reach. Their power is magnified by their access to huge amounts of data about a huge number of people.

Peter recently had a long discussion with a philosopher who'd had considerable interaction with one of the platform companies. They had been at a conference on science and peace at what is geographically the lowest point on earth: a stone's throw from where St. John the Baptist is believed by Christians to have baptized Jesus. On the other side of the Dead Sea was Qumran, where the Dead Sea scrolls written by an ancient Jewish sect had been hidden in a series of caves, not to be discovered until after the Second World War. At night, on the horizon above the hills of the West Bank, were the city lights of Jerusalem—arguably the site of the greatest clashes over beliefs, even though these beliefs evolved from a common core of monotheistic ideas. Somewhere in all of this there seemed to be considerable irony in the conference topic.

But Peter and the philosopher's conversation was focused on the future, not the past or the present. They were discussing the values most attendees at the meeting had taken as fundamental to the human condition, at least in the advanced economies. These values included personal privacy, autonomy, individuals' right to determine what to do with their own lives—in other words, self-determination—and the desire to live in a democracy, where residents have some say over the conditions of their society and where there is a clear governing, but not autocratic, structure to ensure that the commons of society are managed to everyone's benefits. The philosopher challenged these assumptions. He pointed out that these values are not permanent and that the Silicon Valley oligarchs have no particular interest in protecting any of them. Most of these values didn't emerge until the European Enlightenment. Given such transience, he argued, we can't assume that our current core values will be the values of the future. It may be that post-Enlightenment values have proved to be a very successful part of cultural evolution for those who have been privileged by history to live during this period. Every one of these values is affected by the digital revolution, and much could change. We may have to think about what this might mean for various aspects of our lives, such as well-being.

What Does Well-Being Mean?

The International Network for Government Science Advice (INGSA) recently released a report co-authored by Peter that had been prepared at the request of the Organization for Economic Cooperation and Development—the economic think tank of developed nations based in Paris.[13] In this report INGSA considered the meaning of "well-being" in the digital age.[14] The aim was not to decide whether these effects were good or bad, but instead to flag their impacts so that they are more widely acknowledged, especially by the policy community, and can, where appropriate, be managed. It also suggested a research agenda for those actively considering the interface between technology and society. Those involved in the project recognized that well-being was not just about mental health, although that was clearly a key component. There are many dimensions to our individual, social, and civic lives and therefore to our well-being, and many of them are potentially being affected.

For example, in relation to the impacts on us as individuals, the report considered the extent to which the digital transformation is affecting feelings of self-worth, mental health, brain development, opportunities for self-expression and self-actualization, personal health care, privacy, autonomy, and self-sufficiency. When we think about "self," we often think about the quality of our own lives. How do we measure self-worth? Is social media changing our concepts of self-worth from one built around achieving culturally and socially appropriate milestones—which in traditional Western culture would likely be completing education, getting a job, forming a stable romantic partnership, and having a family—to a more narcissistic sense of achievement measured by likes and dislikes on Facebook or numbers of followers on Twitter and by aspirations to "celebrity," based on the metrics of social media? A recent tourist video shown on a domestic flight in Japan exhorted travelers to walk along a particular path, not because of its natural beauty but because photos of it would give more likes on Instagram—so even appreciating nature has shifted from an internal sense of well-being to the quality of the photo we take as seen by others. Even though social media provide new opportunities for freedom of self-expression, they can also lead either to a narcissistically exaggerated belief in one's abilities or to increased self-doubt about one's own worth and ability to achieve external goals. The pressures of social media to portray ourselves in an idealized way can affect how we feel about ourselves and challenge our mental health.

We are increasingly living our lives in public, or in ways that expose us to a very large number of other people. Our sense of privacy seems to be malleable. And what is considered personal and private varies across cultures. Within Western culture, the privacy that was assumed by many to be a fundamental right has been challenged by big data, e-commerce, and the greater engagement with the State. But in many ways we drive the sacrifice of our privacy by using social media and giving in to a self-regarding narcissistic tendency to expose ourselves to public gaze. We willingly let Amazon, Twitter, Facebook, or our cell phone provider know much more than we would wish the State to know about us. Big Brother may be here, but it is not the Big Brother of the State foreseen in George Orwell's novel *Nineteen Eighty-Four,* published in 1949. The Brother who is watching our lives so carefully belongs to e-commerce.

Perhaps the biggest change facing future generations is the pace of change ahead of us. Technologies will continue to emerge, social organization will continue to evolve, financial systems and indeed the characteristics of neoliberal capitalism are undergoing change, the role of the State is clearly changing, and what is certain is that the nature of our life course will change. Most young people will have to acquire a stream of very different skills over their working lives. How will they retrain? How will their working lives evolve? Will stable employment as we know it exist? What will happen to the gig economy? What are the implications for family budgeting, for planning for retirement?

To be resilient in all this uncertainty will require a change in emphasis in skills development. Education has largely focused on knowledge accumulation, but knowledge now resides in bits and bytes. What will be needed are emotional self-control, interpersonal skills, and critical thinking, and these are based, not in the traditional compulsory school years, but before that, in the first few years of life. The greater emphasis needed on early child development will be a challenge for policymakers, educational systems, and parents.[15] The burden of mental illness is now the largest cause of impaired quality of life in many societies. Mental ill-health generally emerges in young people, even if it is not recognized at that age, and affects every part of their lives now and into the future. The digital transformation is clearly a significant factor; but it might also be part of the solution, as evidence suggests that electronic mental health therapies are effective for some.[16]

The internet is changing our social lives in so many ways: in how we conduct our family or romantic relationships, build our social lives and social networks, develop our peer groups and conform with them, contribute to society, and express ourselves culturally. Some of these changes can be healthy, but our value judgments about what is good and bad are themselves being changed by the digital milieu. Because we now live in much more confused and diffuse social networks, generated by the internet and social media, there have been fundamental changes in how we socialize and participate in collective knowledge and opinion formation. We now select groups and information sources that conform to our own established views. No longer do our views mature through debate within our network; instead our networks tend to be defined by our views, which sometimes means being defined in opposition to the views of other networks. As polarization grows, wider social cohesion diminishes.

Our relationship to government is changing. The digital world brings the potential for better access to services, and therefore efficiency in state services—in health, education, justice, taxes, welfare. But the civic discourse is changing. Mainstream media are becoming more fractured, undermined by the loss of advertising revenue to the platform companies, and blog sites and social media have changed the basic forms of social discourse in a society. Certainly not all the responsibility for what is happening can be laid at the feet of the digital gods. Dramatic social and political change happened long before the internet existed. But what appears to be at least quantitatively different is the speed and pervasiveness by which particular views spread and the replacement of civic and reflective comment with polemic and instant responses. Concepts of libel and slander and the ability of the State to protect its citizens have been replaced by enthusiasm for *ad hominem* attacks. Manipulated news, facts, and opinion are much easier to spread. The power of wealth and access to data means that political views can be manipulated through very sophisticated targeting of individual citizens. It is generally considered that such targeted manipulation was a factor in the 2016 US presidential election and in the Brexit referendum. Politicians are being pushed to short-term policies even more than ever, and they are mostly unwilling or unable to control this dynamic.

Digital currencies and high-level encryption have the potential to undermine the ability of the State to manage its own fiscal or monetary policies. Tax bases cannot be ensured in the face of very different employment futures. What might be the future of social safety nets like welfare systems? Personal and public security are increasingly diminished by various forms of virtual, online, and remote surveillance technologies. Surveillance of public spaces might make people feel safer, but in practice it might also limit their personal freedom. Governments have limited ability to protect their citizens or themselves from cyber-attacks, which are giving rise to a continuous digital arms race. A very different world is emerging, and we need an approach that is pragmatic—not dystopian or utopian—to achieve active leadership.

Niche Modification and Mismatch

Throughout this book we have discussed how the interactions between organisms and their environments vary according to their nature and the time scale and degree of environmental change. With every moment of every day we are using our physiology to respond to various stimuli—what we taste, touch, see, smell, and hear. And our thoughts, emotions, and feelings are affected by the interactions we have with other people (as well as with just about everything else in our environment). This is the normal process of homeostasis that we, like other species, have evolved. We have a physiology that can adjust to cope with a wide range of environmental changes. But if the environment changes beyond that evolved homeostatic capacity, we become stressed and our health may be challenged. Changes that are prolonged can threaten our survival or require some other form of adaptation.

As we have seen, there are also other means of adjusting to environmental change. For example, where there is a probability early in life that the organism is going to face a particular kind of environment, its development might adjust to suit that predicted later environment. Recall the Jamaican children developing *in utero* but facing the risk of later malnutrition—they altered their physiology accordingly.

But if the change is great or more prolonged, organisms must have other options to survive. They can migrate or move to a more suitable environment—select a new niche that better matches their biology—or they can modify the environment by constructing a niche, as do termites and beavers, so as to be defended against the ill effects of the change. Humans do that when they use technologies to create forms of protection—such as Inuits' igloos and thick fur clothing.

So organisms have a variety of evolved mechanisms, ranging from homeostasis to niche construction, to deal with changes in their environment. But if the change is sustained, natural selection itself is likely to occur—assuming that there is time for a sufficient number of generations to survive without fatal compromise to the lineage, and that the organism can cope in the interim. But all this assumes that a new equilibrium can be reached between the organism and the environment.

The focus of this book is the distinctive peculiarity of humans, our ingenuity reflected in our continuous and cumulative niche modification driven by technological and social evolution. These have features of "runaway selection." Whereas the peacock's tail may have evolved through runaway sexual selection, ultimately it was constrained by its impact on the survival of male peacocks—they still need to be able to walk, fly, and get into trees to escape predators. In humans, many of the limits that past societies have imposed on social change, often exerted through religious or political authority, have been removed by our recent innovations. Runaway processes may continue until they generate real handicap, whether to our environment, our societal organization, our social interactions, or our mental or physical health. Unlike other species, we do not strive to reach a situation of evolutionary equilibrium with our environment—we continually modify it for reasons other than to promote fitness. So the questions arise: Are our niche-modifying capacities starting to generate serious risks and costs to the human condition? Are they occurring at such a speed that our adaptive mechanisms cannot cope?

We have relied on technologies to make us what we are, to allow us to construct and modify niches as a highly successful species. But we now have to use new technologies to address problems caused by our past technologies. We have seen how the development of antibiotics led inevitably to an arms race with bacteria and the challenge of untreatable bacterial infections. Because we invented industrial-scale agriculture and highly processed foods to meet the demands of changes in our lifestyle and in food distribution systems, we have an epidemic of obesity and its consequences, such as diabetes. So now we have to develop drugs and surgical techniques to cope with the resulting diseases. These drugs and surgical techniques are not without complications, which likely will need to be addressed with new technologies again—and so the cycle continues.

Further, fossil fuels have provided enormous short-term advantages in terms of power production and transport, but now we face the fundamental challenge of global warming and the health consequences of poor air quality. Our modes of communication have changed dramatically, giving us much readier access to our bank accounts, information, entertainment, and friends, but social structures and expectations have changed as a result. Digitalization and its offspring—social media, the internet of things, AI, machine learning—are not only changing our society and social structure, they are altering how we live our lives, and increasingly we observe that these alterations are not always desirable, at least for the future of liberal humanism and liberal democracies.

Some of the questions we increasingly need to ask are: Which technologies might fundamentally harm the human condition as they continue to evolve in a relatively unconstrained way? Which will continue to confer longer-term benefit? Who will define harm or benefit? In reality this will generally require much broader societal discussions of which uses of a technology to encourage and which to try and limit or control, because essentially every technology has both potential benefits and risks. Instead of absolutely banning a technology, we would likely try to define limits for its use. And the word "we" will have different connotations from the perspective of an individual, a community, a society, a national government, or the human species.

We introduced earlier the related concept of precaution. The "precautionary principle" is sometimes misinterpreted to mean that a technology cannot be used until it is proven to be absolutely safe. But that is not the intended meaning of this principle, nor is such a statement scientifically logical. Science can refute a hypothesis, but as the great philosopher of science Karl Popper pointed out, it cannot absolutely prove its truth.[17] So a hypothesis stands until it is refuted. Is a drug absolutely safe because it has been tested on 10,000 people with no reported side effects? Maybe if it had been tested on 10,000,000 people, an adverse side effect would have been found in a few. Does that make it safe, or unsafe? With all innovation we need to make individual and collective assessments of relative risk and benefit. The challenge is that different stakeholders will be affected differently and thus will have very different perceptions of risk. And individual worldviews also affect our perceptions of risk and benefit. The challenge for all levels of societal organization, from family to government, is how to balance these different perceptions.

Are we now confronting situations where some technologies and their consequences are likely to have detrimental effects on the human condition? Are the niches we inhabit no longer modifiable to our advantage, and are they instead being modified in ways that could harm us? Our previous innovation of antibiotics and our ways of using them have resulted in widespread antimicrobial resistance. The success of hygiene has resulted in an epidemic of allergic diseases. The ingenuity of industrial food production is resulting in obesity, diabetes, and heart disease. Our pharmacological creations have resulted in fatal epidemics of opiate and other synthetic brain-altering drug use. And then there is climate change, loss of biodiversity, air pollution, and more. The underlying issues are the same. We have created a whole host of problems, and they often have their worst effects on the sections of our societies least able to respond to them. Now we need to think hard about where we go from here.

Our history over the last 50,000 years is the result of our ingenuity, which drives cultural evolution and in particular the evolution of technologies. Technological evolution accompanied the exponential increase in human population—the two driving each other. Two thousand years ago the world's human population was about 200 million. By 1800 it had grown to about 1 billion, and by 1900 it was about 1.6 billion. By 2000 it had reached over 6 billion, and by 2050 there may be close to 10 billion people on our planet.[18] Technological advances in food production, public health, industrialization, and the development of cities have made this population increase possible, and in turn increasing populations have driven these new technologies.

Technological evolution and social evolution have proceeded at a pace that far outstrips the capacity of biological evolution to respond effectively. So far we have relied on our technological creativity to devise another innovation to adapt successfully to each new challenge we have created. Under some conditions, such as when our ancestors domesticated cows, biological evolution was co-opted to give advantage as well. For humans, however, natural selection is dampened by our ability to control our reproduction, through prevention and treatment of disease and technological protection. Even though we can see some continued evidence of subtle change in our genome that may be due to selection, biological evolution does not offer a solution to our futures.[19]

The question we must ask is whether our more recent technological transformations will have downsides of a scale that will damage the health and future of our species.

Our Biological Future

The Human Genome Project research led to a much better understanding of our genetic makeup. Somewhat paradoxically, though, as we have learned more about genes we have found that our nature is determined by much more than just our genes. Even identical twins can have very different characteristics: they may have very different life stories and experiences or even succumb to different diseases. Science often requires technological advances so it can itself progress.[20] Progress in molecular technologies has led to a focus on the role of epigenetic mechanisms, and in particular the processes by which the environment during development affects the way that genes operate to define the individual characteristics of each of us.

Some people find this disturbing, because they do not like to think that aspects of the world in which they developed have somehow surreptitiously affected their destiny, especially those operating before birth or in infancy, periods for which we have no conscious recall, let alone control. The idea of an epigenetic "memory" of environment set up at this time, potentially affecting behavior, health, disease risk, and many other aspects of later life, can seem spooky. For some, the older idea of our destiny being set by our genes, fixed and predetermined by a long history well beyond our ken, is reassuring and can seem to absolve us of responsibility for what makes us what we are, and possibly for our behavior too.

It is important to understand that inherited genes are not the whole story. It means that we must place much greater emphasis on human development, and on providing the best possible environment for the next generation through attention to early child development and education. The more we understand epigenetic processes, the less tenable it is to hold a fatalistic and orthodox gene-centric view. Far from being occult and disturbing, epigenetic processes are ubiquitous and self-evident. They are continuous and dynamic processes, some of which do not remain static across our lives. This new biology opens up new possibilities for our lives. The expression "life course" has become widely used in biomedical and social sciences in the past twenty years.

The biological realities of genetic evolution make a Darwinian solution to our dilemmas most unlikely. Except following a gross catastrophe, such as substantial climate change, a supervolcanic eruption on the scale of the Toba eruption of 75,000 years ago, a massive meteor impact, or a global pandemic, selection pressures on modern humans are slow, subtle, and indirect. Moreover, as infant and childhood death rates fall and reproduction becomes both safer due to progress in obstetrics and under personal control through contraception, most of us, at least in the developed world, who wish to reproduce can do so successfully. Reproductive technologies have led to new ways for couples to plan their reproductive lives. While there may be subtle selection pressures acting over many generations, they will not address the impact of human activities changing in time frames much shorter than a lifetime. The concept of survival to reproduce as a driver of biological evolution is now essentially meaningless for humans. Our discussion must focus on the realities of the present and the near future and on our cultural evolution.

The impacts of medicine and public health have led to unprecedented increases in the global population. Our population rightly includes many who expect, and deserve, much higher living standards than they have now, with consequent demands for energy, clean water, healthy nutrition, medical care, and meaningful lives. And there are more fundamental changes in our societal structures. At least in the developed world, many children born today can expect to live into their tenth decade, although paradoxically some may live less long than their parents as a result of the impact of obesity and associated noncommunicable diseases.

Social mores have changed too; religion, for many, especially in developed countries, is no longer a unifying framework or source of identity. It has been replaced by multiple identities—political, economic, ideological, recreational, and vocational—and in particular the complex webs of cyberfriends embedded in platforms such as Facebook, Twitter, Instagram, Snapchat, and LinkedIn. Romantic and sexual relationships are, for many, conducted quite differently through matching apps such as Tinder. Family structures are now much more varied. Discrimination based on gender, religion, ethnicity, and sexual orientation may be gradually disappearing, but our digital world has played no small part in giving voice to extreme views and a form of nationalism that is not compatible with dealing with issues of global concern. Intensifying nationalism has the potential to undo much of the progress that has been made on global and regional challenges. As the slowness of the response to climate change suggests, we face the "tragedy of the commons."[21] This tragedy could play out in ways that dramatically change our futures, our lives as social animals.

The explosion of genetic and digital technologies poses new questions that must not be ignored. Genetic modification (GM) is the process, developed in the 1970s, of inserting a whole gene from one species into another. It has played a major role in the production of essential vaccines and medicines. Genes that code for insulin, growth hormone, and many viral antigens are inserted into bacteria and algae, which produce the proteins biosynthetically. And while they remain controversial in some countries, GM crops are now an important part of the world's agricultural and food systems. Gene editing provides an even more precise and specific way of modifying an organism's DNA, using single molecules in a DNA sequence that act as switches to regulate a gene. These can be replaced with surgical precision to change how that gene functions. It is in a sense only doing what nature does through spontaneous mutation and selection, and in that sense it has been seen in some countries as raising fewer ethical concerns than GM. We are already seeing the most accurate of these techniques, called CRISPR / Cas9, being used to produce desirable traits in crops, to produce sheep with short tails and, at least experimentally, to repair damaged genes in human embryos and treat human disease.

These types of new molecular biology raise immense philosophical and ethical considerations. Should molecular biological modifications that cross generations be allowed in humans—for example, to prevent transmission of genetic diseases like Huntington's disease? In 2018 a Chinese scientist claimed to have performed gene editing on human embryos that were reimplanted into the womb, resulting in the birth of two children. The claim has not been verified, and it rightly created a storm of protests based on ethical concerns. Is the genie out of the bottle? If we can do this, what other genetic modifications that are inherited should or should not be allowed? Technologies such as human cloning and genetic modification of human embryos can be regulated legally, and most countries have done so. But will it be possible to universally regulate these newer techniques, which are more automated and cheaper?

Furthermore, some possible applications of these biological technologies can easily cross national boundaries in an uncontrolled manner. For example, the release of synthetic bacteria that might be developed to clean up plastics and oil spills, or mosquitoes that have been manipulated to affect their fertility, will not confine these organisms to any particular geographical area. Further, meiotic gene drive—a form of multiple gene editing that could be used to rapidly push a gene for impaired survival through a population of pests—has been suggested as one way to deal with mosquitos that carry the Zika or dengue virus, but again the modified strains may spread widely.[22] Measures to eradicate mammalian pests can create similar problems, like the imported stoats that are destroying New Zealand's native birds, including the kakapo, a large flightless nocturnal parrot. It will be some time before such technologies have wide application, but the pace of progress in molecular biology puts their potential use just around the corner. Yet we have no processes for having effective transnational conversations on such global issues.

The wisdom of hindsight tells us we should have realized that the development of antibiotics was only the winning of a battle or two in our continued fight against infections, not an all-out victory in the war against them. This is where evolutionary thinking could have been a real advantage. The coevolution of types of bacteria and fungi in environments such as the soil, where they compete for nutrition and space, means that if one species develops a method of suppressing the growth and livelihood of another, then over time that second species will develop a defense that allows it to escape such attacks. Bacteria have mechanisms to transfer DNA from one to another, speeding up this process. So when the penicillin mold secretes penicillin that arrests the growth of a colony of bacteria, eventually there will emerge strains of bacteria that are resistant to penicillin's actions. Such resistance may be rare in the wild, but naturally occurring resistance has been found.[23]

We often treat infections that are not life-threatening with aggressively high doses of antibiotics. And in some countries, small and ineffective doses are given out to placate patients—which does not treat their infections adequately and leads to high levels of antibiotics in the environment. Many infections are viral, not bacterial, and yet antibiotics (which are effective only against bacteria) are often prescribed for these. Worse still is the widespread use of antibiotics in promoting farm animal growth on an industrial scale and sometimes inappropriately in veterinary medicine. Under these conditions the number of interactions between drug and bacteria is greatly heightened and the evolution of resistance is almost inevitable and can be very rapid. Bacteria reproduce at enormous speed—every few minutes, under favorable conditions—and so the potential for a new antibiotic-resistant mutant strain to arise and reproduce is huge.

Strains of bacteria resistant to sulfonamides had appeared by the mid-1940s, and to penicillin by the late 1940s. The same was true for streptomycin by the late 1950s. The twentieth century saw enormous progress in the development of new classes of antibiotics, each one offering new hope by targeting aspects of bacterial metabolism, reproduction, or invasion which offered chinks in bacterial defenses. Such drugs were developed about every five to ten years up until about the 1970s. But for almost every class of drug, resistant strains of bacteria have developed within just a few years.

Our cultural evolution of technologies to defeat the bacteria that occur naturally in our world has led to our being mismatched to that world—because inevitably the bugs will fight back. We are now in a position where, as badly as we need them, it will take considerable time and money to find or develop new classes of antibiotic agents, if indeed we can. The tit for tat—or perhaps the better analogy is arms race—of drug development, then the emergence of drug resistance, then development of a new drug, then the development of resistance to the new drug, is causing considerable concern. Experts such as Dame Sally Davis, England's chief medical officer, see antimicrobial resistance as a rapidly emergent major threat to human health. So-called superbugs have emerged which are almost unkillable. There is an international movement to reduce the use of antibiotics. Doctors are advised not to prescribe them for colds or minor infections from which the patient will recover naturally in a few days or weeks, usually developing their own immunity. There are protocols in place to combat the development of antibiotic-resistant strains of organisms in hospitals, where the battle between organisms and drugs is fought most intensely every day. And greater attempts are being made to address their overuse in agriculture. But just as with all innovations, once they are established we have to bear the bad consequences along with the good. There is never any going back. We cannot reverse the presence of, for example, penicillin-resistant strains of bacteria; all we can do is restrict our use of penicillin and work to find other ways of defeating the bacteria. There is no certainty that we will be able to win in this arms race; antimicrobial resistance is here to stay, and even our ingenuity may not be able to fix it.

Climate change highlights our difficulties. The evidence for anthropogenic climate change is overwhelming, and the likely pattern of consequences has been well-described. Yet debate continues, fueled largely by misinformation, itself driven often by vested interests. The global climate system is complex and thus there remain uncertainties as to the precise trajectory ahead. But, despite such uncertainties inherent in any predictive science, the risks to our planet are certain and alarming. Yet concerted action escapes us. There are short-term partial solutions in the switch from fossil fuels to renewable energy and in changed consumer behaviors, but long-term solutions largely depend on as yet unproven and unscalable carbon capture technologies. And what if they fail? Would we have to find some geoengineering solutions and, if so, what global process would we have to agree to in order to deploy them? This is where our ingenuity has brought us.

Our Digital Future

Managing biological technologies may be easy in comparison with the most pervasive of our new technologies—the digital world. The former largely involves individual decisions about technology use or applications for which regulation is possible, although there are concerns about biological technology's potential for "dual-use" for malevolent purposes or for accidental consequences from "backyard" use. But digital technologies have emerged in a very different way. Their applications and the ever-expanding suites of derivative technologies have been driven largely by the private sector, boosted by security and intelligence interests and, in particular, by countries that see the digital technologies as the driver of economic productivity and competitiveness.

The internet has been a tremendous boon to communication and information transfer. But now it dominates our lives. Where would we be without email, Amazon, and Google, to name the obvious resources? But it is much more pervasive than that. Our public utilities and services—telephone, power, water, waste disposal, hospitals, taxation, street lighting, policing, medical records, and so forth—all depend to an increasing extent on the digital world. Without GPS, how would we know where we are and, more importantly, how would our cars and our planes?

As we discussed in Chapter 7, we are being naive if we do not honestly confront some of the spillover costs of these developing technologies. The internet does not discriminate between reliable and unreliable claims. Combined with the "echo chambers" of social media that it has created, it has fueled the creation of a new environment—one in which access to information is seen to be sufficient by itself and expert interpretation of that information is no longer needed. This is the "post-expert" world that has received so much prominence in the United Kingdom with Brexit, and in North American and European politics with populist movements. The internet permits the rapid spread of information and misinformation. So alongside its empowering contributions to knowledge access, it has fueled the accelerated distribution of misinformation for multiple purposes, from political to commercial. This is the "post-truth" pandemic. In a sense post-truth is nothing new—manipulative misinformation has been common since societies first developed hierarchies of control for political, commercial, or egotistical purposes—but the internet has made it so much easier and so much more pervasive.

Think how selective we are in what we look at and do on the internet. The cognitive biases inherent in our very nature as social animals mean that for the most part we hear and see only what we want to hear and see.[24] Experiments in the behavioral sciences have shown that values, rather than facts, largely determine what we believe and how we make decisions.[25] Economists once assumed that all decision making has a rational basis, and it has taken them a long time to recognize that a large portion of our decisions are based on emotions and biases. Modern politics demonstrates how easy it is for collective decision making to ignore realities, especially those that require long-term thinking.

There are important value judgments here. Some people see only benefits arising from developments in digital technologies, with minimal downsides. Others, especially those in government, are so focused on the potential short-term impacts on economic productivity that they do not want to consider broader societal issues: staying ahead of competitors in productivity is more important to them. Still others see real dangers ahead and can be alarmist—the futurist and controversial entrepreneur Elon Musk, for example, sees AI as heralding an apocalypse that could end humanity.[26]

Many Western governments are now focused on the implications of new digital technologies for the future of work. As robots equipped with machine learning capability and AI enter the production arena, what will happen to jobs? Some studies suggest that up to 30 or 40 percent of them will be severely affected, either disappearing altogether or more likely being changed so substantially that they will require a different skill set, which many workers will not be able to learn.[27] Will the new jobs suit the workers whose former jobs were eliminated by the new jobs? And will they be in the same location? It is likely that there will be a very difficult transition. Many commentators suggest that the decline in industrial employment in the American Midwest that was at the heart of recent political change originated not so much in shifting trade patterns but instead in new technologies and the emergence of robotic heavy manufacturing.

We can already see early examples of what the future may hold. There are already AI services sold to law firms to take over a range of routine roles previously played by lawyers—what will the lawyers do instead? Apps such as Babylon in the UK are already taking over the traditional roles of some health professionals—so will we need as many doctors and nurses? Most experts think the roles of humans in the future economy will depend more on emotional than cognitive skills—on skills that are developed early in life and require very different approaches to education. But there are broader implications. Fewer people may have long-term employment. Will this be good or bad for our mental health? What will it do to a country's tax base and the ability for the government to support and provide services to its citizens?

Pragmatism

We are neither techno-optimists nor techno-pessimists. Instead we take a realistic and a pragmatic view. There have been enormous changes in the human condition as a result of digital technologies—many have been for the good, although they raise important questions about our future direction of travel as the digital revolution continues to rapidly modify our niche. But the revolution has some particular characteristics. It is market-driven by the power of the technology, the information it holds, and the concentration of power and influence it brings. We quickly see its convenience but do not have time to reflect on its wider consequences. Because it is so cleverly marketed and so pervasive, new editions of the technology emerge very quickly and without the kinds of societal discourse that have been associated with technological developments such as nuclear power or GM crops.

Technological innovation resulting from our ingenuity is at the core of what makes us what we are, but all technologies have costs. Given the unique pervasiveness and impact of this accelerating wave of innovation, we do not think these difficult issues can or should be avoided. What is needed is a pragmatic acknowledgment that this technology is on an almost unstoppable trajectory that will have major impacts on the human condition. As societies, we need to find the appropriate balance between its development by entrepreneurs, its use, and its regulation. Can we shape the trajectory to emphasize its benefits and minimize its downsides? Problems of regulation are made much more difficult by the transjurisdictional nature of the innovations and the speed with which they are produced. The responses by society and policymakers have been relatively ineffective, delayed, and reactive. In general there is a lack of enthusiasm and commitment to consider governance and ethics of the digital world as a societal issue. This is the problem with runaway cultural evolution. The history of evolution is littered with examples of species that have become extinct, usually due to failure to adapt to environmental change or to compete with rival species. Are we similarly at risk as a result of our own accelerating technologies?

Reality Matters

Our confidence about our perceptions of the world around us is at risk, as are our concepts of reality. The infamous phrase "alternative facts" is increasingly being used without embarrassment—many people have come to think that reality and truth are whatever they would like them to be, not something that needs to be supported by evidence. Political opinion polls show that many people are not concerned by this as long as the alternate facts meet their own biases and interests. Many apps allow us to manipulate our images to give an enhanced impression of how we look or where we are. And augmented reality is only in the early days of its possible applications. The internet provides huge resources of information on any subject we choose to explore, and we usually have no easy way to assess such information, in terms of who provided it and how reliable or unbiased it is. The best we can hope for is that reviews available online will give us a clue, but similar problems apply to these—and the use of bots to bias ratings or opinion rankings of everything from hotels to clinical services makes the problem even worse. Fact-checking sites cannot keep up with the tsunami of misinformation. When we are searching for information on something we feel is important, perhaps a health problem or a financial issue, our uncertainty about which sources to trust can be a real concern.

If manipulations of reality become more common, how can a child's developing mind know what is true or real and what is fake? The fundamental basis of our perceptions about reality, which we depend upon for our learning as we develop, seems challenged. Knowing that nothing can be trusted or that nothing is necessarily how it seems might be important, but it does not offer an alternative solution. Reality testing is key to mental health. What will be the effects on the newest generation of infants, for whom exposure to the new media is as commonplace as building bricks or teddy bears were for former generations?

While we increasingly understand through insights from both the natural and the social sciences where we have come from, and what makes us what we are now, we are left to ponder the core question of this book. Our ingenious technology and social innovation has allowed us to be a very successful species, but can it continue to keep us successful? Or is it fundamentally changing us and our world in ways that are beyond our control? Has the pace of technological innovation become so fast that we will not be able to cope with the inevitable disadvantages it will bring? We need to be more conscious of these issues if we are to shape our futures for the best.

9 | FUTURE

The mountain gorillas we met at the beginning of this book are living essentially as they did 100,000 years ago. They still live in groups with a dominant silverback, several females, perhaps some submissive other males, and offspring of different ages. They eat the same meals of bamboo and other leaves they have always eaten, except that they have recently taken a liking to the bark of eucalyptus trees, which were imported about a hundred years ago from Australia and now grow prolifically around the edge of the park. Their use of sticks as tools has not changed in millennia. Until recently, the equilibrium they had reached between their environment and their evolved biology allowed them to survive in their niche. Sadly, because of the explosion of human settlement in this crowded and very fertile part of Africa, their habitat has been restricted and their numbers are threatened by poaching. There are perhaps only 800 or so mountain gorillas left.[1] Now their survival depends on us, on the very active efforts by national parks, scientists, and the Gorilla Doctor service.[2]

How different the history of our own species has been. We have changed our habitats and our societies dramatically as we have spread from Africa to every part of the globe. Our population has expanded exponentially, and the lives we live, while highly variable in context and custom, demonstrate how we have continued to modify the niches we inhabit and our material culture. We have invented, embraced, and exploited technology built on technology. Our evolution has become dominated by the processes of cultural rather than biological evolution. The drive to modify our environments, the niches we inhabit, has not come from just our Darwinian fitness—our need to survive to reproduce. We have used our ingenuity to modify our niches for many other reasons, such as for pleasure, for profit, or for politics.

This book has offered examples of the mismatches that emerge as the pace of cultural evolution far exceeds any biological evolutionary processes. Humans are unique among all species in their capacities for such rapid and cumulative cultural evolution. It appears that as we evolved into large-brained, bipedal hominins living in small social groups—and with hands that allowed unique dexterity—the combination of group knowledge, the capacity to teach and learn, and the ability to communicate thought and instruction through language fueled our remarkable evolutionary path. But now our trajectory is one of runaway cultural evolution.

We are more than niche constructors: we did not end up living in stereotypic self-made environments, like the termite or the beaver. Rather, depending on where we lived and particularly in recent times, we made progressive and increasingly extensive modifications to our environments. When those modifications had downsides, we resorted to further technological innovations to cope with the consequences of earlier developments. We are creative innovators, so these were much more than just "duct tape" patches, temporary fixes; they were innovations of real ingenuity. But as our settlements grew, newer social and technical solutions were needed, because close-knit living raised new versions of old issues involving food supply, waste disposal, spread of infectious disease, transport, communication, and social organization.

Now we wonder if our species can indefinitely rely on ingenuity to cope with the adverse consequences of our irrepressible capacities to innovate and build technologies. The issue of climate change highlights this concern. Its origins in fossil-fuel-based technologies, expanding populations, and increased deforestation are not in doubt for scientists. Excessive carbon dioxide and other greenhouse gases in the atmosphere almost certainly will lead to unacceptably high global temperatures. Technologies will have to be deployed on a massive scale to protect us and the planet. Whether through new forms of renewable energy, nuclear power, carbon capture technologies, or the use of genetically modified crops, it is essential that we make changes to many of the ways we live. But we face the problem of a potential "tragedy of the commons," for global governance is weak—and compromised by the self-interest of the powerful. Science is ignored, denigrated, and misconstrued to deny that the problem exists or to misrepresent the evidence for it. Will we even have to pursue the ultimate in speculative technologies—geo-engineering, such as methods to reflect or block more of the sun's rays—to keep ourselves, and other species on our planet, in existence? The fact that this question can even be seriously asked indicates the perils to which we have exposed ourselves and our planet by not adequately considering the consequences of our ingenuity—of the relentless drive to technological innovation that has become our own nature.

Antimicrobial resistance raises similar issues. No doubt we will find new antibiotics as we explore soils and plants for new chemical agents, but the history of these drugs is clear: bugs evolve quickly, and resistance appears and spreads fast. Will the reduction in infectious disease as a major killer in economically advanced countries be seen in the future to have been just a transient interlude in the relationship between humans and microbiota? Who will win this arms race?

The invention of high-energy, tasty, processed foods changed eating behaviors in much of the world. But this change came at the cost of obesity, skyrocketing rates of diabetes and heart disease, and so the need for more technological fixes—new drugs, bariatric surgery, and so forth—but even so, some scientists predict a fall in life expectancy in the next generation.[3] Some governments are sufficiently worried that they are introducing taxes on sugar, against the objections of the powerful food industry, of course. Taxes on foods that promote obesity might help to address the problem, but they will hardly solve it.

Our current era of digital transformation may transform the human condition on a scale not seen since the development of agriculture and settlements. Not one aspect of human existence is untouched by this transformation; it affects our cognitive function, our psychological and emotional health, our relationships with those close to us and with whom we work and play, how we work, where we go, and what we do. Market capitalism, the powers of national governments, and even the nature of the relationship between state and citizen are being altered.

We Have Choices

We face some fundamental questions. Will we wait passively to see where technology takes us? Or do we want to be more deliberate in the controls we apply to it—and, if so, how do we do this individually and collectively? How do we get the best from technology and restrict its worst effects?[4] What are the respective responsibilities of individuals, elected governments, international nongovernmental organizations, and global corporations?

We have looked at our past through an evolutionary lens, and it seems logical that we must look to the future through the same lens. Evolution in itself has no goal—it is not directional. This is certainly true for biological evolution, as genetic change is largely a matter of variation and selection simply to fit an existing environment. Biological evolution, as the great evolutionary biologist John Maynard Smith reflected, is the outcome of various transitions that open up more possibilities.[5] Disruption in biology is met by migration, adaptation, or extinction. Survival sets the rules of biological evolution. The human species, however, has another option: further cultural innovation. Indeed, the extreme dream of the transhumanists is to escape the limits of mind and body altogether. Even if we cannot predict the future, we can foresee the range of scenarios that might emerge and perhaps summon the ingenuity to shape them.

Shaping Our Social Lives

As discussed previously, our brains evolved for living in social groups on the order of 150 people, and group identity and membership have played a critical role in making us what we are. But social structures are changing as we interact with each other in bigger groups, and societies are becoming more diverse and less cohesive. The impact of the internet and social media has clearly expanded both the number of people with whom we link and the size and number of the groups of which we feel a part. What used to be a limited network of friendships has now become a vast array of acquaintances. This might confer real benefits: the internet provides new opportunities to gain supportive and helpful information and contacts with others who are experiencing similar challenges. Consulting a virtual doctor about a sensitive health problem may involve less embarrassment and judgment than seeing a family doctor, and it certainly gives an "answer" in minutes rather than days or even weeks. But there is also growing suspicion that this increasingly digitalized world is having detrimental effects on mental health, especially in young people.

The absence of direct face-to-face contact, especially if there is no kinship, may mean that the quality of support afforded by virtual friends is less robust, reliable, or stable. The authority and guidance of parents or teachers may be replaced by validation of behavior or attitudes from social media networks and virtual games. It should be noted that different personalities may be affected in different ways by their use of the internet. Sweeping generalizations must therefore be replaced by deeper understanding of vulnerabilities and resilience at both the group level and the individual level. Further, some societies and cultures are more individualistic and others are more collective in their assessments of well-being, so we might also expect cultural differences in how people are affected by the digital world.

Social support networks traditionally operate on the basis of shared experiences, validated by personal contact. Personal contact helps on multiple levels to establish and enrich relationships. The motives for virtual friendship may be very different, and they may reinforce dimensions that are not necessarily supportive. Peer support and pressure may have different meaning and impact when one is exposed to a large network, which never meets in person, as opposed to encounters on the village green. Too often we resort to email when we want to avoid what might be difficult or embarrassing personal interactions. With the ability to be anonymous on the internet, it is easier to be unpleasant online than in person, and on the internet some people will say and do things that they would never do in person. Sociologists and psychologists have known for a long time that peer groups themselves can be damaging if they establish unhealthy expectations and behavior—this has been suggested as a critical factor in such things as pressure on young people to have sex.[6] The online digital world has greatly intensified various sorts of harms that groups or individuals within groups can do to individuals. Cyberbullying within peer groups has been linked to youth suicides.[7]

Perhaps it is no surprise that in societies where more people are connected online than ever before, loneliness and isolation are increasingly common problems. And yet, for others the internet is a critical part of their ability to avoid isolation.

Of course, access to smartphones and the internet is not universal; even in high-income countries, some portions of the population, mostly the poor, are not online. This can only exacerbate the inequities and inequalities that have increased in these countries, especially since the global financial crisis of 2008. And age matters, too. In the past, most societies saw elders as sources of wisdom gained through experience. Now, instead, in person, ageism often replaces respect; and online, older people also have less influence if they are less comfortable using new technologies. Many older people feel that their experience has simply become irrelevant to younger people, who expect the internet to deliver all the knowledge they need. This might seem like a temporary divide between "digital native" young generations and older ones who grew up in a pre-digital world. But innovation is exploding at such a fast pace that we should expect future generations similarly to be challenged in their later years by technologies well beyond those that existed in their formative years.

Shaping Our Institutions

As humans came to live in increasingly larger conglomerations, the social institutions and structures that empowered this transition evolved. Simple hunter-gatherer groups were held together by kinship and the need to work collectively. The development of agriculture led to people living together in settlements, with more structured divisions of labor. Formal organizational hierarchies, such as priesthoods and monarchies, arose as means to organize populations. Linked to this was the development of bonds and rituals of group identity based on religion or loyalty to elites. Hierarchical societies were the outcome.

Democracy ameliorates some of the worst effects of hierarchical society by creating accountability for those in power to those being governed. But it seems that current cultural, technological evolution is challenging democracy itself. While democracy has never been perfect, the digital world exacerbates its problems. It further undermines the provision of unbiased information to citizens upon which they can make choices; it makes the provision of false and manipulated information easier; it puts manipulative power into the hands of those who have access to means of targeting information to large numbers of citizens; it replaces opportunities for reasoned discussion with a myriad of opportunities simply to react, click, and vent; and it encourages identification with extreme views. The rise of populist movements and the manipulation of electoral processes are very worrying. Many institutions have been compromised or at risk of being so. Financial, social, legal, and political institutions are all being affected by current technological transformations, and the influence and power of some private-sector organizations, operating beyond national jurisdictions, challenges the sovereignty of national governments.

The news media increasingly pander to the preferences and prejudices of their audiences, which may in turn have been sculpted by social media manipulation. With captured and preselected audiences, coverage of issues and events becomes weak on analysis and strong on entertainment or incitement value. One way this is achieved is by presenting strongly held opposing views without regard for how representative they are. For example, residents of the United Kingdom who are in favor of fox-hunting with dogs, and those who are not in favor of it, might be given equal column inches or airtime, irrespective of how many people there actually hunt foxes with dogs nowadays. This results in political discussions, at national or local levels, that are increasingly oppositional, personal, and confrontational—all because this is how the media have decided to vie for our attention.

Are the days of meaningful discussions about policy and principles over? The dynamics and the echo chambers of social media and our preselected newsfeeds reinforce our prior biases and drive a more polarized form of politics and discourse. Not all of the problems of democracy can be attributed to media's digital transformation, but the point remains that we need institutions that better support our need to live in cohesive groups, despite their being large and multidimensional. Our efforts to change culture have had impacts on the very institutions we created to enable us to live and thrive in groups.

Shaping Public Policy

The modern, so-called Westphalian system of nation-states with clear borders and implied contracts between citizens and governments contrasts with an earlier model of feudal lords battling their neighbors for power and resources and exploiting their serfs as soldiers, food producers, and payers of taxes or tithes.[8] But the Westphalian model is still relatively young and somewhat fragile. Feudal systems are reemerging in failed states such as Afghanistan and Yemen, and increasingly autocratic regimes are reemerging in many countries—even in some that had previously embraced democracy.

Representative democracy emerged as the form of government to which Western culture, and increasingly global culture, aspired. But the quality of dialogue that democracy was intended to promote is being replaced by more populist and extreme forms of argumentation engendered by the digital infosphere. Digital communication facilitates immediate interaction between citizens and politicians, and in many ways this is desirable, but it shifts the expectation to making decisions for the short-term instead of the longer-term. There is a growing sense that the increasingly short-term focus of public policy is limiting long-term strategy development on the very issues that will affect us most in the future.

Policymaking is inevitably a complex business where any decision made, any choice, including inaction, affects multiple stakeholders in different ways. There are always trade-offs, and, in principle at least, policymakers and politicians must weigh the likely impacts of their decisions on different interest groups. Furthermore, policymaking is inevitably messy, with multiple actors, formal and informal, elected and unelected, contributing to the process. And because there are always options, how government officials formulate the alternatives and present them to the politicians who ultimately decide on the course of action—or inaction—becomes very important. The reality is that policymaking will always be heavily influenced by a series of value-laden factors: public opinion, electoral contract and ideology, fiscal and diplomatic considerations, and so forth. Concepts of electoral and actual risk and benefit inform these decisions.[9]

We need to think more about how policymakers interact with those who inform their understanding. Where do the evidence and data produced by scientists fit in? Science, broadly defined, cannot generally create policy, but it can assist policymakers by shedding more light on the dynamics of the system in question, thus providing a clearer understanding of the interventions available and their possible impacts. Better outcomes from the policymaking process are likely when there are mechanisms to ensure that scientific evidence adequately informs it, and when policymaking systems are designed to remain adaptive and flexible as technologies evolve.[10]

Shaping Citizens

One of the most important aspects of all our lives is the education we gain through both early-life and lifelong learning opportunities. It may be through formal channels—preschool, primary and secondary school, college, apprenticeships—or it may be by less formal mechanisms such as clubs, networks, and community-based activities. Nowadays most education is likely to be supported by digital technology, but for reasons discussed above, it is not appropriate for learning to be delivered by this means alone. As the Brazilian educator Paulo Freire observed many years ago, learning is an active process, not simply a matter of "banking" information in a passive recipient mind.[11] Teaching therefore has to be a transactional process rather than just a transmission of information. The transactional aspect is essential to enabling students to challenge their situations in life, which they must learn to do if they are to play their parts as active citizens of a better world. More recent pedagogical theorists have extended this thinking to argue that teaching must be approached as an intellectually disruptive and subversive activity if it is to instill inquiry skills in learners and encourage them to think for themselves rather than mindlessly accept received ideas. We believe this is more important in the digital age than ever before.

Another feature of good teaching is that the student is encouraged to learn from mistakes. This requires a supportive, nonjudgmental environment, where trial and error are assumed to be part of the learning process, and failures are valued for the knowledge they can yield. This is hardly the environment of the digital world, which is quick to condemn every misstep and has an indelible memory of past activity. It is possible now for every site you have visited, every purchase you have made, and every person you have contacted to be noted, recalled, and communicated to others, without your permission.

Good teaching helps instill resilience in the learner. Resilience enables children to have the confidence to take on new challenges, however daunting or unexpected, without expecting to be defeated by them. We cannot foresee the situations young people will face in even a few decades, but we can educate them to be resilient as their world presents its novel threats and opportunities.

But there may be an even more critical role here for early education. The workshop that led up to INGSA's report on well-being in the digital age, discussed in Chapter 8, argued that the skills that will be most critical to the digital age are those of self-control, resilience, empathy, and emotional intelligence, and that these skills have their foundations in the first few years of life. The argument was that the pace of change in society is so continuous and fast that without emotional resilience the impacts on mental health of those growing up in this changing environment may be severe. Furthermore, as automation and AI displace many traditional jobs, and as job structures change rapidly with technological innovation, it is likely that skills such as empathy, emotional, and social intelligence will be the basis of much future employment. As serial retraining becomes more likely, passage through the job market and society itself will become more difficult without these so-called noncognitive executive functions. It is very clear that children who grow up in less than optimal environments have deficits in these skills, and such deficits are difficult to remediate later. The report argues that governments should take a greater interest in the quality of early childhood environments and the development of social skills.[12]

Through the Lens of Science

Science stands apart from other knowledge systems in that it is defined by attempting to remove preconceptions and bias from the processes of collection and analysis of information about the world around and within us. It is not immune from value judgment, because our choices of what to study and how to study it are never value-free. And there are always judgments about the quality and sufficiency of data on which to draw conclusions. Much science fails to meet these standards for many reasons—but without science we are left to form our judgments about our world based on mere belief, dogma, tradition, and anecdote.

Good science is a set of processes. These processes vary according to the discipline and the question being asked. Without these, the knowledge generated cannot be distinguished from mere belief and dogma. The processes of science are designed to eliminate bias as much as possible in the collection and analysis of the data. But scientists themselves are humans with biases, values, and worldviews that influence what they do and how they do it. Good science circumvents these biases by using the tools of experimental design, critical analysis, repetition, statistics, correction, and peer review—all of which are inherent to the scientific tradition. However, whether rigorously acquired scientific data and discoveries are taken up or not by society depends on the society's value-laden choices.

Science itself is a form of human culture that has evolved. The scientific method was known in the Arabic world 2,000 years ago.[13] It was rediscovered in Europe by Sir Francis Bacon in 1620. In 1660 the Royal Society for the Advancement of Knowledge, one of the world's most prestigious scientific bodies, was founded; it is the oldest continuously active scientific academy. Science as we now know it had started to flourish at that time. Sir Isaac Newton was the seventeenth person to sign its charter book—a priceless parchment folio that continues to be signed with a quill pen by every newly elected Fellow. The Royal Society started the first scientific publication in 1665, *Philosophical Transactions,* which is still published. Pascal, Harvey,

Humboldt, Buffon, and other luminaries started to produce large amounts of knowledge and ideas in various domains helped by technological inventions like the microscope and by the increasing voyages of discovery by proto-colonial powers. Philanthropy by rich and royal patrons was the mainstay of such science.

When countries began to want to understand their natural resources (and others' resources) better, governments started to invest in science and as the institution of the research university developed in Germany in the nineteenth century, science started to take its modern form. Across two world wars, the manifest advantages gained through scientific and technological innovation convinced governments that greater funding of science must be prioritized as a core strategic investment. In the late twentieth century, many governments' research initiatives were spurred by Vannevar Bush's landmark 1945 report to the US president, *Science, the Endless Frontier*, which argued that science could do for peacetime development even more than it had done in wartime.[14]

Science in the postwar period was initially dominated by reductionist approaches, focusing on very specific and narrow questions. With the development of accessible computation from the 1970s, much more complex and systems-based science became possible. Statistical approaches could be applied in ways that were not possible previously and that permitted much more complex interactions to be investigated and evaluated. Ecological and environmental science came of age. The explosion in biological sciences in the decades following the discovery of RNA and DNA ushered in a new wave of medical science. It became possible to study the processes underlying disease in much greater detail and to develop drugs to target mechanisms rather than just treat symptoms.

So-called systems approaches operate on the premise that a more complete, albeit never absolutely complete, understanding of a phenomenon—a chronic disease, changes in patterns of weather, brain function—can be gained by integrating knowledge across various fields and levels of organization. Systems approaches recognize the reality of complexity and that systems are rarely closed—that is, rarely unaffected by external influences. They also recognize all sorts of complex feedback loops and interactions. Human societies are open systems—as are ecological, climate, and most biological systems—so their workings cannot be understood without taking external influences into account. Growing computational power has allowed systems approaches to become increasingly important in advancing knowledge, in part by making it possible to analyze complex, nonlinear interactions and dynamics beyond first-order effects.[15]

Scientists have claimed that greater investment in science will drive greater returns to society, and many governments have responded positively to this. But funders increasingly expect credible evidence that their investments are producing valuable returns. As the need to demonstrate the impact of public investments intensifies, the returns are often couched in economic terms. However, it is important to also consider whether the science involved affects public policy, informs public health or environmental strategies, creates diplomatic advantage, supports security, or improves the resilience of a society against natural disasters (including earthquakes, floods, and pandemics). Many people, one hopes, also value the autotelic returns on scientific investment, such as the enhancement of a country's reputation when its basic research in, say, astronomy or particle physics produces insights that fill in gaps in fundamental knowledge.

As the scientific enterprise has grown, its isolation from society has started to break down. Science is about explaining what we know and what we do not, acknowledging the complexity in every subject we study. And the very areas where governments and citizens have most interest are those where there are often many unknowns, uncertainties, and complexities. This does not mean science cannot assist—indeed, science is essential to shaping our decisions about the future.

Because of their very nature, the social sciences, public health, environmental science, and economics command high levels of public interest. These sciences, however, have their applications in areas where there are many value-based perspectives within society, and these are often in dispute. For example, vested interests have clashed over how much priority should be assigned to research on climate change, or gun control, or embryonic stem cell research. This type of scientific problem—where science and societal values have to engage—is sometimes called *post-normal* science.[16] It concerns areas of science where there are currently many unknowns and there will always be residual unknowns, and where decisions about implication and actions are urgent, but where the science involved interacts with public values that are themselves in dispute. This is the realm in which much of the science that informs policy today operates. Indeed, we can see that virtually all the conundrums about the future of the human condition fit within this definition.

Scholars of post-normal science point out the importance of real dialogue between scientists and the rest of the community at the earliest stages in a research program. This is not the proud science communication of new discovery; instead, it is engaging the community at the very beginning of the initiative and then continuing to do so. This means involving interested parties who are not scientists in developing what research to do, how to do it, and what to do with the results. These are challenging concepts for the institutions of science, but given the issues that lie ahead, they cannot be considered to be of only theoretical interest.

Experimental Societies

By definition, humans have always lived in an experimental mode—a mode in which our existence as social animals is the experimental unit. As we have progressed as a species, every social innovation, every technological development, has involved taking a step into uncertainty. Such experiments have perhaps been accidental rather than intentional as we developed agriculture and settlements and evolved our social systems as we lived in progressively larger aggregations. But adaptive adjustment was possible because of a much slower rate of innovation. Today we are living with constantly developing innovations that have substantial impact, and the consequences of our niche-modifying activities are becoming increasingly apparent. This rapid rate of change is creating manifest pressures. Addressing these issues will require us to apply our ingenuity to develop new technologies and innovations, if our species is to continue to thrive. But it will also require reflection on how to channel our ingenuity, and, in some cases, where to restrain it. We must make choices about what technological innovations to pursue. In the balance hang our well-being, our social relationships, our health, our environment, our economies, our governments, and our planet.

When societies were small and the pace of innovation was slow, there were inherent constraints on our social and technological evolution. But with growth in knowledge, development of new technologies, and massive advances in science, the pace of innovation has reached the steepest part of the exponential curve. New devices constantly appear; we now simply expect a relentless output of new apps, new drugs, and new models of every machine from the car to the carpet cleaner to the space capsule. The same is true in terms of social innovation, where new trends driven by social media can infest populations very quickly. Think how quickly Pokémon Go emerged, and then faded, as a worldwide fad.[17]

When the adoption or use of innovation is constrained, the constraint is often based on grounds other than scientific. Stem cell research and the associated developments in regenerative medicine have been inhibited in the United States for ethical, religious, and political rather than scientific reasons. The choice not to vaccinate a child is driven by anecdote and inappropriate knowledge of risk rather than science. The objections to GM foods are primarily based on philosophy and ideology rather than on science. The choice to stop nuclear power in Germany was political rather than technological.

Proportionate societal constraints are appropriate, for every technology has a potential downside. There is arguably no modern technology that has not been considered, or indeed used, to make weapons or to do intentional harm; cyberwarfare, autonomous weapons, and even trucks driven at speed at pedestrians are some of the new battle tools. But we need to be willing to explore when constraint is appropriate and when it is not.

As we discussed in Chapter 8, acceptance or rejection of a technology depends on our individual and collective views about risk and precaution. This in turn is affected by our prior biases and worldviews and those of our peer group. Most often we see technologies in terms of short-term direct benefit as opposed to long-term collective benefit (which we tend not to personalize). The argument of extreme precaution can be used to inhibit innovation—this may be justified or it may be used to advance a political goal. Different stakeholders may have different views of such justifications, as in the case of GM crops.

Innovation is not possible without accepting some uncertainty, and science can never give absolute certainty. The challenge is how to put all this in perspective. Should we use nuclear power to reduce the use of fossil fuels? Should we use genetic techniques in crops to reduce greenhouse gas emissions from agriculture? Hard questions need to be discussed without falling into the sloganeering that inhibits productive discourse and yet is so popular on social media.

The combination of AI and big data offers enormous opportunities to understand more about ourselves and our planet and to make better choices. Several countries are trying to make productive use of such potential. New Zealand, for instance, is using its Integrated Data Infrastructure initiative to link up much of its government agencies' data on health, education, justice, the welfare system, earnings, and so forth into a unified data system. The aim is to understand the causes of disadvantage and find better ways to intervene. But there are serious questions—about social license, privacy, how the data will be used—that cannot be ignored. Governments must be transparent in how they curate, secure, and use their social data. Citizens also have to understand that they live within a social context and that social pooling of personal data can help further social progress. But equally, access to data can allow governments to have greater control over their citizens. At the same time, citizens have perhaps unwittingly given enormous control to the platform mega-companies who can use the enormous amount of data they have on each of us, and the metadata they obtain by combining data about us as individuals, to determine what we know and influence how and what we think. In a microcosm, we can see the complex interplays between technology, society, and policy and how perceptions of risk and benefit need to be worked through.

Looking Globally

In 2015 the leaders of 194 countries, meeting at the United Nations, agreed upon seventeen Sustainable Development Goals (SDGs) that they hoped could be attained by 2030.[18] Those goals include eradicating poverty, promoting gender equality, preventing hunger and ensuring healthy nutrition, reducing disease, addressing climate change, protecting the oceans, and promoting better governance. Many of the SDGs are highly aspirational and provide a sense of direction rather than operationally feasible goals. They have not been constructed in a way that links easily to domestic political drivers and institutions. They do not align well with the way governments operate, and they do not provide an understanding of the inevitable trade-offs and consequences of any specific action. Some of the goals, and the 169 more-specific targets they include, could pull policy in opposite directions if addressed simultaneously. And in general, governments want to be seen to respond to the wishes of their citizens rather than to the aspirations of a global agency such as the United Nations. But flawed or not, they are a positive statement about where our political leaders claim they want to take us.

The breadth of issues illustrated in the SDGs can be viewed through a variety of lenses. Even though all seventeen goals were accepted by all UN members, a skeptic might point out that the agenda does not create obligations or commitments; governments can interpret objectives in their own terms, and their voluntary reports back to the UN can be based on their own performance indicators rather than pre-agreed international targets or metrics. On the other hand, the SDGs highlight the many issues that we must confront—all of which arise because of the ways in which our species has evolved to constantly modify the niches we occupy across the planet. They create a framework for international, regional, and local action, not just by governments but by everyone.

Our technologies have allowed us to live everywhere, have facilitated population growth, and have in general improved the lives of many. Yet, as this book has emphasized repeatedly, we must look hard at ourselves and understand that there may be limits to our ability to change nature. Yet we—the global population of this planet—now face many potentially dire consequences of our past innovations, and they can be solved only by looking at the long-term and not by grabbing at temporary, short-term fixes. This is easy to say, but few want to sacrifice their interests for others whom they do not perceive as being in their in-group. This is particularly evident in the debates over climate change, where progress has been stymied by national and sector interests. Public discourse and policymakers need to find ways through this dilemma. Obviously there is some irony in the push to develop more technologies to solve the very problems that past technologies have created, but that is inevitably how we will address these issues. The challenge that our society's digital transformation highlights is that we must bring together civil society, the private sector, and policymakers in the effort to map a path ahead. Together we must find ways to shape the direction in which technology will take us.

Most of these issues require an understanding of both national interests and technological change. Few countries have effective domestic systems for integrating technological and scientific thinking and foresight into their policymaking processes. Ideology still largely trumps analysis and evidence. Virtually no country has a holistic approach to thinking about the digital transformation and the multitude of nontechnical issues that arise from it. But many of the issues will require transnational action to be effectively tackled. This is clearly the case for the digital transformation and for solving the problems of ocean acidification and pollution, climate change, water, food and energy security, antimicrobial resistance, pandemic risk, and so on.

Consider, as a thought experiment, the range of problems that would emerge if geo-engineering were the only way to deal with climate change. We have no established processes for deliberating about how, when, and whether to undertake adopting such technologies. We need to create these processes. And for some forms of life-science technology, we need more transnational approaches. Mosquitos do not need passports, viruses and bacteria (whether naturally occurring or synthetic) do not stay within artificial borders. Perceptions of precaution, risk, and benefit vary enormously within and between nations, and increasingly their citizens are influenced more by opinions on social media than by science. Yet we need to find the way ahead.

Society, its leaders, and especially the younger generation must engage with these issues. There are already examples of our failure to do so. Our slow progress in addressing climate change is a challenge for which future generations are likely to pay a high price. The application of genetic modification to crop agriculture—which, most scientists believe, holds promise and will almost inevitably play a role in our futures—has led to claim and counterclaim, use of misleading data, and inappropriate platforms for debate, all of which polarize views rather than move us toward a consensus. But at least there has been some debate over climate change. In contrast, the escalating predominance of and transformations of digital technologies, which have much more direct and immediate impacts on our lives, have been introduced and employed without discussion or oversight. Part of this discrepancy lies in different stakeholders' varying perceptions of cost and benefit, and of risk (which, somewhat ironically, are heavily influenced by the dominance of digital social media). This is inevitable, and a well-functioning democracy should be able to handle such differences in viewpoints. But democracy is not well served by increasingly polemical and confrontational dialogue. Irrespective of our fundamental beliefs and worldviews, we cannot avoid collectively confronting our future. If we don't want to be doomed to a Spencerian future determined by a frightening form of social Darwinism where those who control the technology have all the power, we must rapidly develop a broad consensus on how to address some core issues.[19] The Agenda 2030 and the SDGs are an attempt to achieve that, but time is pressing.

Finding the Way Ahead

There is an old saying: "If you do not know where you are going, any road will do." To a large extent this is true of the human condition. Evolution has no predetermined direction. As the great evolutionary biologist and writer Stephen Jay Gould pointed out, if we could replay the history of evolution, it would almost certainly lead in a very different direction.[20] We probably would not be here.

This unpredictability is even more true of cultural evolution—we only have to think how quickly fashions change. Further, there are fewer constraints on cultural evolution than on biological evolution, and the pace of technological change has both predictable and unpredictable elements. What is clear is that the human condition is being affected, swiftly, in ways that challenge our biology, behavior, and social structures. At the same time the very technologies producing these challenges also have enormous value to us. This is the paradox of our using ingenuity to change nature by niche-modifying strategies.

Regarding the future, should we be depressed or excited? Optimistic or pessimistic? The best approach is to be realistic and pragmatic. It is inevitable that humans will continue to develop technologies, but just as we have done with many technologies already, we need to put boundaries of use around them—even though scientists, technologists, businesspeople, governments, and ordinary citizens will disagree over where the boundaries should be. We need processes that assist us to reach consensus on such matters but such processes must be adaptable to changing circumstances. We are inevitably going to have to use technology to address the problems we have created, and we must review risk assessments as we learn more. This requires much more sophisticated dialogues than those concerned simply with short-term political expediency. We need to take one of those forms of ingenuity that evolution has given us—namely, the ability to be self-reflective—and synthesize what we can from the multiple disciplines of science and the humanities to understand our nature and apply our ingenuity wisely.

We should think carefully about what we need to be happening in our education systems. For instance, how can young children be taught critical thinking? And such considerations do not only apply to school. Too many natural scientists graduate from university with little or no knowledge of the social sciences, of the relationship between science and society, of scientific philosophy, or how science interacts with policy. And social scientists need to understand the contribution of natural sciences. Scientists also need to consider other epistemologies and how to interact with them—for example, traditional and local knowledge are valuable resources for many societies and cannot be ignored as we seek ways to enhance societal development in all parts of the world.

A far more integrated approach to technological development is needed. Policymakers and societies need to consider what science, broadly defined, can tell us about where we have come from and where we are going. It will be important to find ways to have critical conversations about the limits and uses of technology. Such conversations are inherently difficult, but made more so by the current state of the internet and social media.

We fall into a trap when we base our understandings on reductionist thinking. When we do so, either in science or in public policy, there may be a temporary sense of progress but it soon stalls. Culture wars have hurt progress in both areas of life. Too much of public policy is based on ideology, personal bias, and short-termism. Because the digital world favors greater polarization of views, protecting centrist politics poses a challenge. But what is crucial is that the evidence used to assist policymaking should be robustly developed. It must be reliable and multidisciplinary. It must be presented in a way that sums up what science knows and does not know, and what is reasonable to conclude, acknowledging that there are always uncertainties. It must point out the realistic, pragmatic, and promising options to the policymakers.

Policymakers have the challenge of knowing when to be *laissez-faire* and when to be proactive in response to new technologies. And they need help not only from experts but from a better informed and a more engaged public. We need to find mechanisms for conducting these difficult but essential conversations if we are to protect human life and our planet in the future.

Evolutionary thinking is necessary. We have thrived as a species, and thrived increasingly well as we learned how to change nature, for 150,000 years or more.[21] But there may be limits to how much we can modify our world, because nature always responds in return, changing our nature, too. Evolution is a fact, not merely a hypothesis, and it is the central unifying concept in biology. By extension, it affects almost all other fields of knowledge, and must be considered one of the most influential concepts in Western thought. Understanding this is fundamental to our biology, our social structure, and society. Thinking carefully about our strategies as a consequence will ultimately determine whether our children can continue to thrive.

Our Nature

In 1937 to 1938, the eminent British physiologist Sir Charles Sherrington delivered a series of lectures in Edinburgh entitled *Man on His Nature.*[22] Europe was about to descend into another war, so it was an opportune time for Sherrington, a scientist who had also read widely in philosophy and history, to provide his view of what these events revealed about human nature. The achievements of science in the first part of the twentieth century, especially in embryology and genetics, held the prospect of explaining our biology in greater depth than ever before, even perhaps reducing biology to the sciences of chemistry and physics, which at that time was thought to be the ultimate goal. The benefits of such reductionism seemed to be just around the corner.

In his last lecture in the series, "Conflict with Nature," Sherrington referred to the work of the eighteenth-century thinker and scientist Joseph Priestley. Writing in 1768, twenty years before another traumatic event in Europe, the French Revolution, Priestley had held out great hopes for the innovations that new scientific work could bring. Sherrington quoted Priestley's view of a future "glorious and paradisiacal beyond our imagination. . . . Nature, including both its materials and its laws, will be more at our command; men will make their situation in the world abundantly more easy and comfortable; they will probably prolong their existence in it, and will daily grow more happy."[23] Sherrington remarked that Priestley's prediction was correct, except for the last part. Indeed, whatever the triumphs of science, human happiness in Sherrington's world was just about to be drastically threatened.

Sherrington decided not to say anything about the conflict of men against men, preferring to end his lecture series, in the gendered language of his time, by noting the paradox that "Man" as a species seems always in conflict with Nature; but yet, having evolved consciousness, Man is able to understand how his own nature, his "urge to live" as a predatory animal, conflicts with the altruism necessary to live in the world, not only as a social animal but respecting the existence of other species. Sherrington acknowledged that the idea of Man as "master of his fate" was challenging: "Rather, it would sometimes seem to him, he is merely a tragic detail in a manifold which goes its way without being even conscious of him. . . . Master of his fate? Around him torrential oceans of energy; and his own energy by comparison a drop which trickles down the window-pane." It is a sobering metaphor. Sherrington emphasized, however, that human nature drives us to change the planet for our own purposes, even though we know we are not its only guests. He also remarked that, as the process continues, we too are changing.

Developmental physiologists spend their time thinking about the effects of environment in the sense in which it is often considered a part of "nurture." They research the environment provided before birth by the placenta and the womb; they study parents' bodies, diets, and lifestyles; they investigate whether the environment before birth matched that encountered later. And they consider the influence of all this on how the little plastic organism that was each of us, once upon a time, acquired its unique set of characteristics. There is plenty of scope for changes in human nature there—just as Sherrington had indicated in his lecture "The Wisdom of the Body," in which he described the sequential development of the embryo.[24]

Our observations about how the rapid technological developments during the young lives of the "millennials"—internet, smartphones, fast food, Facebook, and so forth—shaped them, and shaped even more the lives of the subsequent "Generation Z," convinces us that the battle between humans and their nature is still very much raging. The tacit assumption is that, because these brilliant new technologies are the products of human ingenuity, humans are winning. But the more we look, the more it appears that humanity is, at the same time, losing—whether we turn our attention to the epidemic of obesity and its association with morbid disease, to cyberbullying and mental health, to climate change, to antimicrobial-resistant bacteria, or to societal organization, loss of social cohesion in some societies, and conflict. All these result from brilliant technological ingenuity, yet all challenge fundamental aspects of human biology.

The often unintended impacts of ingenuity are a fundamental part of the human story. And because of the game of survival in evolution, changing nature through our technologies is part of human nature. In turn, the world that we collectively create through niche modification changes us. Once we recognize this, we can begin to see our world and our nature—and our ongoing development, which so powerfully mediates between the two—in new ways. We can begin to suggest some real solutions; diagnosis must precede effective treatment. At the same time, we can recognize that this is only the beginning of the next phase of our evolution, which will take all our ingenuity to bring about.

NOTES

Introduction

1. Charles R. Darwin, *Notebook B:* [Transmutation of species (1837–1838)], Darwin Online, ed. John van Wyhe, http://darwin-online.org.uk/content/frameset?pageseq=1&itemID=CUL-DAR121.-&viewtype=side.
2. Darwin kept to himself his lack of belief in theistic creation, but he was frank in his essay "Recollections of the Development of My Mind and Character," which he wrote for his family. After his death this was published as an autobiography, edited by Darwin's son, who removed several passages about Darwin's views on God and Christianity. These were later restored in an edition published on the hundredth anniversary of the publication of *The Origin of Species.* See Charles R. Darwin, *The Autobiography of Charles Darwin, 1809–1882: With the Original Omissions Restored* (London: Collins, 1958), edited and with appendix and notes by his granddaughter Nora Barlow.
3. Carl Zimmer, "When Darwin Met Another Ape," *National Geographic,* April 21, 2015, https://www.nationalgeographic.com/science/phenomena/2015/04/21/when-darwin-met-another-ape/.
4. Richard Owen (1808–1892), a distinguished anatomist and doyen of British science, was a staunch creationist and took every chance to undermine Darwin's evolutionary arguments. In 1857, in a lecture at the Royal Institution, he claimed that the structure of the gorilla brain was so distinct from that of the human brain that it argued for distinct creation. Thomas Huxley (1825–1895) disagreed with some of Darwin's ideas, at least in their early development, but all the same became Darwin's "bulldog." He and Joseph Hooker (1817–1911), one of Darwin's other freethinking naturalist colleagues, emerged as leaders of post-theological British science. Huxley was

merciless in his attack on Owen in his own lecture at the Royal Institution in 1858.

Chapter 1 | The Outback

1. Termite colony structures differ depending on the environment and the species of termite. For example, in well-drained sites the species *Amitermes laurenis* builds mounds that are small and dome-shaped, but in seasonally flooded flats its mounds are large and oriented north–south. Mounds built by *Nasutitermes triodiae* can reach up to eight meters high and rather resemble cathedrals, after which they are named. In proportion to their body size, termites construct the largest structures of any living creatures.
2. Robert Edmond Grant was one of Darwin's mentors in Edinburgh; he introduced Darwin to the study of marine invertebrates. He was a freethinker and early evolutionist, and a strong supporter of the work and theories of the French transmutationists Etienne Geoffroy Saint-Hilaire and Jean-Baptiste Lamarck. But he went beyond Lamarck in accepting a common origin for plants and animals. He later became a professor of comparative anatomy at the newly formed University College London, where his radical political beliefs and convictions about evolution put him into conflict with one of Darwin's critics, Richard Owen.
3. Charles Lyell (1797–1875) was the most influential geologist of his era and was largely responsible for a shift in understanding of how earth's features were formed gradually rather than through catastrophic events. He had a strong influence on Darwin and became a close friend, although he remained a creationist until arguably the tenth edition of his *Principles of Geology*. He first met Darwin soon after Darwin returned from the *Beagle* voyage. Adam Sedgwick (1797–1875) was another distinguished geologist and one of Darwin's lecturers at Cambridge. Darwin accompanied him on a geological expedition to Wales prior to joining the *Beagle*. Sedgwick, an Anglican reverend, remained friends with Darwin although he was a firm creationist and strongly opposed to Darwin's theories. John Stevens Henslow (1796–1861) was one of Darwin's tutors at Cambridge and particularly inspired him on botanical matters. Henslow left Cambridge a few years after Darwin to be a priest in a country parish, but he remained a mentor to Darwin and played a central role in Darwin's being invited to accompany the ship's captain on the *Beagle*.

4. Alexander von Humboldt (1769–1859) was a Prussian naturalist, explorer, and polymath who was perhaps the most notable European explorer-naturalist at the beginning of the nineteenth century. In 1799–1802 he made a scientific exploration of South America, and his *Personal Narrative* describing those years was one of the books Darwin took with him on the *Beagle*.

5. Robert FitzRoy (1805–1865) was born into an aristocratic family and served as an officer on the first voyage of HMS *Beagle* to survey Tierra del Fuego for the British navy. After its commander committed suicide, FitzRoy was appointed captain. On that trip he took Jeremy Button and three other Fuegians as hostages after a dinghy was stolen, and he decided to return them to England to be "civilized." FitzRoy commanded the *Beagle* on its second, more famous voyage (1831–1836), on which Charles Darwin sailed—as did the three surviving Fuegians, who were to be returned. FitzRoy was famous for his temper, but by and large he and Darwin got on well—except when the issue of slavery came up. In 1839 FitzRoy published a multivolume account of the voyage, entitled *Narrative of the Surveying Voyages of HMS Adventure and Beagle*—which included Darwin's *Journal and Remarks, 1832–1836* (now known as the *Voyage of the Beagle*) as its third volume. FitzRoy later became the second governor of New Zealand and had an unhappy time trying to manage tensions over land sales between British settlers and the Māori; he increased rather than reduced tensions, which led to his recall. In 1851 Darwin supported his election to the Fellowship of the Royal Society. In 1854 FitzRoy became the first head of a department to develop marine weather data that in time became the Meteorological Office. He popularized the use of barometers, stressed the importance of weather prediction for sailors, coined the term *weather forecast,* and set up storm-warning systems. When *The Origin of Species* was published in 1859, FitzRoy was distressed by the role he had played in the development of Darwin's ideas, and at the famous 1860 Oxford evolution debate between Thomas Huxley and Samuel Wilberforce he spoke in pain from the floor. FitzRoy's longstanding troubles with depression led him to end his own life.

6. William Paley (1743–1805) was a British clergyman and philosopher. His book *The Principles of Moral and Political Philosophy* (1785) was a core component of examination requirements when Darwin was a student at Cambridge University, which at that time was essentially part of the Anglican Church.

Paley's book *Natural Theology or Evidence of the Existence and Attributes of the Deity* (1802) was an early influence on Darwin, who had a copy of it on the *Beagle*. In it Paley advanced the argument that complex biological design is evidence of active deistic creation, famously using the analogy of the watchmaker. This argument has continued to be advanced by intelligent-design and creationist movements, which prompted Richard Dawkins to give the title *The Blind Watchmaker* (Harlow: Longman, 1986) to the book in which he refutes such arguments. Darwin had clearly rejected Paley's arguments by the time the *Beagle* returned to Britain.

7. These notes included a 35-page "sketch," prepared in 1842, summarizing his ideas and their supporting arguments and evidence, which Darwin intended as a personal note; and, from 1844, a 189-page manuscript drafting his thoughts on how evolution could—and has—occurred.

8. Wallace had spent several years in Southeast Asia collecting and documenting thousands of specimens, before arriving at the island of Ternate in present-day Indonesia in 1858. Soon after moving into his house, he fell severely ill and was mostly confined to bed. Despite this—or possibly because of it—Wallace experienced a *eureka* moment as his thoughts and observations consolidated to formulate the idea of natural selection. After his recovery, he immediately committed his ideas to paper, sending the essay to Darwin, with whom he had previously corresponded. Darwin was by then close to finishing his own tome on the same ideas, but this was now at risk of being overtaken. Darwin's scientific colleagues Joseph Hooker and Charles Lyell therefore organized a joint presentation of both Wallace's and Darwin's work at the Linnaean Society less than a fortnight later, on July 1, 1858.

9. For an extensive discussion of Wallace's perceptions of the relative contributions made by himself and Darwin, see John van Wyhe, *Dispelling the Darkness: Voyage in the Malay Archipelago and the Discovery of Evolution by Wallace and Darwin* (Singapore: World Science, 2013).

10. Janet Browne, *Charles Darwin: Voyaging* (London: Pimlico, 2003).

11. The first use of the word *evolve* in the modern biological sense was likely by Robert Jameson, a professor of natural history in Edinburgh and founder of the Plinian Society, a freethinking (that is, not bound by the religious dogma of the day) natural science club of which Darwin was a member. In an

1826 paper Jameson wrote that Lamarck had "explained how higher animals had *evolved* from the simplest worms." See Adrian Desmond and James R. Moore, *Darwin* (New York: Norton, 1992).

12. The final paragraph of the first edition of *The Origin of Species* (1859) reads:

> It is interesting to contemplate a tangled bank, clothed with many plants of many kinds, with birds singing on the bushes, with various insects flitting about, and with worms crawling through the damp earth, and to reflect that these elaborately constructed forms, so different from each other, and dependent upon each other in so complex a manner, have all been produced by laws acting around us. These laws, taken in the largest sense, being Growth with reproduction; Inheritance which is almost implied by reproduction; Variability from the indirect and direct action of the conditions of life, and from use and disuse; a Ratio of Increase so high as to lead to a Struggle for Life, and as a consequence to Natural Selection, entailing Divergence of Character and the Extinction of less improved forms. Thus, from the war of nature, from famine and death, the most exalted object which we are capable of conceiving, namely, the production of the higher animals, directly follows. There is grandeur in this view of life, with its several powers, having been originally breathed into a few forms or into one; and that, whilst this planet has gone cycling on according to the fixed law of gravity, from so simple a beginning endless forms most beautiful and most wonderful have been, and are being evolved.

Darwin expressed little interest in the issue of the ultimate origin of life, his focus being what followed from the first life-form. In public, but not in private, he was careful in how he dealt with issues of religion. In later editions of *The Origin of Species* (there were six editions in all), he amended the last sentence to read: "been originally breathed by the Creator into a few forms." This amendment was likely made to reduce public debate and it certainly gave comfort to his wife, Emma, who was a believing Christian.

13. Erasmus Darwin was a physician and philosopher who, in his book *Zoonomia* (1794), which Charles had read, described life as a continuous chain from the most primitive forms to humans. Later, while in medical school,

Charles was under the mentorship of the zoologist Robert Grant, who was an ardent supporter of Lamarck's transmutational ideas.

14. Jean-Baptiste Lamarck (1744–1829) was a Parisian naturalist who arguably was the first to present a coherent view of transmutation, arguing for a progression from simple to more-complex forms and for the importance of organisms' adapting to their environment. He came to believe that the use or disuse of a physical feature could lead to morphological changes that could be inherited. The phrase "inheritance of acquired characteristics" is often referred to derisively as "Lamarckism," but in fact Lamarck never used this phrase, and his contributions to science are much broader. He introduced the term *biology* as we understand it. The concept of use-inheritance was widespread among his contemporary thinkers.

15. Mendel published his work on inheritance in 1866—seven years after the publication of *The Origin of Species*—in an Austrian natural history journal, where it languished in obscurity and remained untranslated from the German. In 1900 three European botanists independently confirmed Mendel's findings, prompting the English biologist William Bateson (who would later coin the term *genetics* to describe the study of heredity) to publish his book *Mendel's Principles of Heredity* (1913) supporting Mendel's work. In 1905 the Danish botanist Wilhelm Johannsen wrote the pathbreaking book *Arvelighedslærens elementer* (The elements of heredity), which introduced the concepts of gene, genotype, and phenotype.

16. Jonathan Mark's book *What It Means to Be 98% Chimpanzee: Apes, People, and Their Genes* (Oakland: University of California Press, 2002) is an insightful discussion of the meaning of genetic differences between species and within species, and the implications for racism.

17. Jack Cross, *Great Central State: The Foundation of the Northern Territory* (Kent Town, S. Aust.: Wakefield Press, 2011).

18. Also known as the mulga snake, this highly venomous species can grow up to 2.5 meters long and deliver as much as 150 mg of venom per bite—the largest known output of all snakes worldwide. Its venom is cytotoxic, myotoxic, and neurotoxic.

19. Recent research suggests that humans first arrived in Australia 65,000 years ago, about 10,000 to 20,000 years earlier than had previously been thought.

Chris Clarkson, Zenobia Jacobs, Ben Marwick, et al., "Human Occupation of Northern Australia by 65,000 Years Ago," *Nature* 547 (2017): 306–310. It can be assumed that the journey was planned, given that the group included enough men and women to start a new population. However, the means by which they managed to traverse the ocean separating Asia and Oceania remains a mystery, as does their motive for doing so. It has been speculated that the appearance of land bridges during times of low sea levels may have played a role, but even at the lowest sea levels, significant water crossings would have been necessary.

20. When Europeans first established a colony in Australia, in the late eighteenth century, the population was organized into about 700 groups, which spoke a total of about 250 languages. Computer modeling of genetic and linguistic data has helped to map when and how humans spread over the continent.

21. Allison L. Perry, Paula J. Low, Jim R. Ellis, and John D. Reynolds, "Climate Change and Distribution Shifts in Marine Fishes," *Science* 308 (2005): 1912–1915.

22. Niche construction is the process by which organisms change the environments they inhabit (niches) through their metabolism, activities, and behaviors, thus influencing evolutionary processes.

23. Beak characteristics differ according to diet. For example, the long, pointed beaks of cactus finches are suited to retrieving seeds from cactus fruit, whereas the short and stout beaks of ground finches help with foraging for seeds on the ground. These finches have played a pivotal role in our understanding of evolution. They have since also been a useful model to demonstrate and study important evolutionary concepts underlying species emergence.

24. S. van der Kaars, G. H. Miller, C. S. M. Turney, et al., "Humans rather than Climate the Primary Cause of Pleistocene Megafauna Extinction in Australia," *Nature Communications* 8 (2017): 14142.

25. Australian Institute of Health and Welfare, *Australia's Health 2016,* Australia's Health Series no. 15, cat. no. AUS 199 (Canberra: AIHW, 2016).

26. Aristotle came up with the concept that would later be phrased in this way—*Natura abhorret vacuum*—by the fifteenth-century French writer François Rabelais.

27. The migratory route undertaken by humans in spreading from Africa to the Australian continent on the other side of the globe is still the subject of much debate. One postulate is that an early wave of migrants crossed the Red Sea from Ethiopia to arrive at the Arabian Peninsula, and then traveled along the coasts of South and Southeast Asia to eventually reach Australia. A second migration wave, thousands of years later, led to the population of Europe and Asia. In 2016 a set of studies that independently analyzed the complete genomes of various populations concluded that Australians diverged from Eurasians after splitting from Africans; this indicated that Australians and other non-Africans were all descendants from a single migration wave out of Africa, and thus ruled out an earlier, southern migration route. See Luca Pagani, Daniel John Lawson, Evelyn Jagoda, et al., "Genomic Analyses Inform on Migration Events during the Peopling of Eurasia," *Nature* 538 (2016): 238–242. However, more-recent analyses have provided support for the "southern route" proposal. See Nicholas G. Crawford, Derek E. Kelly, Matthew E. B. Hansen, et al., "Loci Associated with Skin Pigmentation Identified in African Populations," *Science* 358 (2017): eaan8433. In this study, more than 1,500 African individuals were measured for their precise skin tone and then genotyped, revealing four genes significantly associated with skin-color variation. Global populations were then analyzed for their possession of these variants, which allowed an estimation of the migratory pattern out of Africa. The study showed that variants found in individuals with the darkest skin were also found in native Australians, Melanesians, and South Asians—precisely the groups that would have inherited the mutations from migrants following the southern route.
28. A. W. C. Yuen and N. G. Jablonski, "Vitamin D in the Evolution of Human Skin Color," *Medical Hypotheses* 74 (2010): 39–44.
29. N. G. Jablonski and G. Chaplin, "Human Skin Pigmentation as an Adaptation to UV Radiation," *Proceedings of the National Academy of Sciences* 107, suppl. 2 (2010): 8962–8968.
30. Darwin suffered from chronic ill-health of a very uncertain nature for the rest of his life after his voyage on HMS *Beagle.* The illness afflicted him with vomiting, weakness, and frailty and inhibited his writing and research quite severely. He attempted numerous cures in spas, particularly cold-water treatments. This illness is a dominant feature in many Darwin biographies. For

example: Desmond and Moore, *Darwin;* and Janet Browne, *Charles Darwin: The Power of Place* (Princeton, NJ: Princeton University Press, 2003).

31. Darwin landed at the Galápagos Islands in 1835, toward the end of his famous voyage around the globe. The finches he observed there varied so markedly in beak shape and size across the different islands that he initially mistook them to be members of different families. Back in Britain, the ornithologist John Gould noted that the bird specimens were in fact all species of finches, prompting Darwin to focus further on his ideas of transmutation of species and of natural selection as a key evolutionary process. Darwin does not refer to the finches as support for his theory in *The Origin of Species.* However, a mythology has grown up around the role of the Galápagos finches in his thinking. For an extensive discussion of how this scientific myth grew, see Frank J. Sulloway, "Darwin and His Finches: The Evolution of a Legend," *Journal of the History of Biology* 15 (1982): 1–53.

32. Peter Raymond Grant and B. Rosemary Grant are a husband-and-wife team of distinguished evolutionary biologists who have spent many years studying natural selection within finch populations on Daphne Major, one of the Galápagos Islands. They have been able to show that natural selection on beak size can occur within a single generation. Their work is summarized in many scientific papers and in several books, including *Evolutionary Dynamics of a Natural Population: Large Cactus Finch of the Galápagos* (Chicago: University of Chicago Press, 1989); *How and Why Species Multiply: The Radiation of Darwin's Finches* (Princeton, NJ: Princeton University Press, 2011); and *40 Years of Evolution: Darwin's Finches on Daphne Major Island* (Princeton, NJ: Princeton University Press, 2014).

33. Michael K. Skinner, Carlos Guerrero-Bosagna, M. Muksitul Haque, et al., "Epigenetics and the Evolution of Darwin's Finches," *Genome Biology and Evolution* 6 (2014): 1972–1989.

34. Swanne P. Gordon, David Reznick, Jeff D. Arendt, et al., "Selection Analysis on the Rapid Evolution of a Secondary Sexual Trait," *Proceedings of the Royal Society B: Biological Sciences* 282 (2015), https://royalsocietypublishing.org/doi/abs/10.1098/rspb.2015.1244.

35. Y. E. Stuart, T. S. Campbell, P. A. Hohenlohe, et al., "Rapid Evolution of a Native Species following Invasion by a Congener," *Science* 346 (2014): 463–466.

36. Patrik Karell, Kari Ahola, Teuvo Karstinen, et al., "Climate Change Drives Microevolution in a Wild Bird," *Nature Communications* 2 (2011): 208.

37. Ryan P. Kovach, Anthony J. Gharrett, and David A. Tallmon, "Genetic Change for Earlier Migration Timing in a Pink Salmon Population," *Proceedings of the Royal Society B: Biological Sciences* 279 (2012): 3870–3878.

38. Ying Song, Stefan Endepols, Nicole Klemann, et al., "Adaptive Introgression of Anticoagulant Rodent Poison Resistance by Hybridization between Old World Mice," *Current Biology* 21 (2011): 1296–1301.

39. To compensate for lower oxygen levels at higher altitudes, most people—including the Aymara people—have elevated hemoglobin levels. This, however, is not seen in Tibetans, who have normal hemoglobin concentrations equivalent to those of Americans who dwell at sea level. Cynthia M. Beall, Gary M. Brittenham, Kingman P. Strohl, et al., "Hemoglobin Concentration of High-Altitude Tibetans and Bolivian Aymara," *American Journal of Physical Anthropology* 106 (1998): 385–400. Genetic studies revealed that Tibetans are more likely to possess variants in a gene involved in production of red blood cells, thus accounting for their relatively lower hemoglobin levels. Cynthia M. Beall, Gianpiero L. Cavalleri, Libin Deng, et al., "Natural Selection on EPAS1 (HIF2α) Associated with Low Hemoglobin Concentration in Tibetan Highlanders," *Proceedings of the National Academy of Sciences* 107 (2010): 11459–11464. This population also demonstrates other physiological adaptations not observed in the Aymara, such as lower arterial oxygen levels, higher resting ventilation, and an absence of pulmonary vasoconstriction.

40. Matteo Fumagalli, Ida Moltke, Niels Grarup, et al., "Greenlandic Inuit Show Genetic Signatures of Diet and Climate Adaptation," *Science* 349 (2015): 1343–1347.

41. The omega-3 fatty acids, especially EPA and DHA, are essential to our diet and health. This was discovered many years ago when it was realized that the Inuit have very low levels of cardiovascular disease despite consuming a diet very high in fat. These fatty acids originate in plants, and the correct balance between omega-3 and omega-6 fatty acids is critical to our ability to make cell membranes of the right composition, and also to produce chemical agents that control a range of body functions, from platelet

stickiness to the function of our small blood vessels. At some point in our evolutionary past we lost the ability to synthesize these compounds, so we are dependent on receiving an adequate supply of them in our diet. Low body levels of omega-3 fatty acids are associated with cardiovascular disease and, even more strikingly, also with the chances of, for example, having a second heart attack in those who have had one such attack. The simplest way to gain adequate intake of omega-3 fatty acids is to consume one or two portions of oily fish such as salmon, tuna, or herring every week. The salmon gain these fatty acids from the smaller fish they eat, which acquire them from plant sources. But salmon will eat all sorts of other things, so farmed salmon are often fed plant material or cereals directly. This reduces the amount of omega-3 fatty acids in their bodies, making them less healthy as a food source for us. This innovation is driven by cost, because farming large numbers of salmon is cheaper than catching them in the wild, and feeding them on plant oils rather than fish oils is cheaper, too.

42. *Genetic drift* refers to changes in the genetic structure of a population independent of selection. Genetic drift most often occurs when a small founder population with little genetic variation becomes isolated from its parent population. For example, the first peoples of the Americas are almost exclusively blood type O, none having blood type B, and very few having blood type A. This suggests that the founder population, which likely crossed from Siberia to Alaska over a land bridge in the ice age perhaps 14,000 years ago, contained no members with blood types A or B, or at least no members who left progeny. Benito Estrada-Mena, F. Javier Estrada, Raúl Ulloa-Arvizu, et al., "Blood Group O Alleles in Native Americans: Implications in the Peopling of the Americas," *American Journal of Physical Anthropology* 142 (2010): 85–94.

43. B. F. Voight, S. Kudaravalli, X. Wen, et al., "A Map of Recent Positive Selection in the Human Genome," *PLOS Biology* 4 (2006): 0446–0458; J. Hawks, E. T. Wang, G. M. Cochran, et al., "Recent Acceleration of Human Adaptive Evolution," *Proceedings of the National Academy of Sciences* 104 (2007): 20753–20758; Yuan-Chun Ding, Han-Chang Chi, Deborah L. Grady, et al., "Evidence of Positive Selection Acting at the Human Dopamine Receptor D4 Gene Locus," *Proceedings of the National Academy of Sciences* 99 (2002): 309–314; K. S. Pollard, S. R. Salama, N. Lambert, et al., "An RNA Gene

Expressed during Cortical Development Evolved Rapidly in Humans," *Nature* 443 (2006): 167–172; C. Wills, "Rapid Recent Human Evolution and the Accumulation of Balanced Genetic Polymorphisms," *High Altitude Medicine & Biology* 12 (2011): 149–155; Joshua M. Akey, Michael A. Eberle, Mark J. Rieder, et al., "Population History and Natural Selection Shape Patterns of Genetic Variation in 132 Genes," *PLOS Biology* 2 (2004): e286; Felix C. Tropf, Gert Stulp, Nicola Barban, et al., "Human Fertility, Molecular Genetics, and Natural Selection in Modern Societies," *PLOS One* 10 (2015): e0126821; Jaleal S. Sanjak, Julia Sidorenko, Matthew R. Robinson, et al., "Evidence of Directional and Stabilizing Selection in Contemporary Humans," *Proceedings of the National Academy of Sciences* 115 (2018): 151–156.

44. Jerome C. Rose and Richard D. Roblee, "Origins of Dental Crowding and Malocclusions: An Anthropological Perspective," *Compendium of Continuing Education in Dentistry* 30 (2009): 292–300; Peter Gluckman, Alan Beedle, Tatjana Buklijas, et al., *Principles of Evolutionary Medicine,* 2nd ed. (Oxford: Oxford University Press, 2016). Another example is increased exposure to artificial light indoors and prolonged close-range optical focusing, likely contributing to increased rates of myopia, which is especially prevalent in some populations that also appear to have a strong genetic predisposition to myopia. Ian Morgan and Kathryn Rose, "How Genetic Is School Myopia?," *Progress in Retinal and Eye Research* 24 (2005): 1–38.

Chapter 2 | Survival

1. Darwin's orchid, *Angraecum sesquipedale,* is an epiphyte found in Madagascar. Darwin made many botanical studies, including of orchids and how they were fertilized by insects. After being sent several flowers, Darwin noted the extremely long spur of *A. sesquipedale.* He surmised, after a number of experiments in which he tried to remove pollen from deep in the spur, that there must be a pollinating insect with a proboscis long enough to reach the nectar at the end of the spur. He recorded this in his book *On the Various Contrivances by Which British and Foreign Orchids Are Fertilized by Insects, and on the Good Effects of Intercrossing* (1862). The idea that there could be an insect with a thirty-five-centimeter proboscis was ridiculed, and such a species was not believed to exist. Alfred Russel Wallace entered the debate a few years

later with a paper setting out in detail a sequence through which the moth and the flower could have coevolved with no guidance other than natural selection. The debate ended when Darwin's prediction was confirmed with the discovery of such a moth in Madagascar in 1903.

2. Darwin focused much thinking, inquiry, and experimentation on the issue of how plants and animals reached islands, and to the distinct speciation that followed reaching different territories. He came to the view that this was the action, not of distinct creation, but of accidental dispersal by wind, rafts, and birds. He sought evidence from numerous correspondents and had debates with Hooker about whether seeds could survive in salt water. Then, in 1855, he conducted experiments at Down House in which he explored seeds' ability to survive in salt water and then germinate. He enlisted many people to help him and published announcements in the *Gardeners' Chronicle*. He explored whether birds' feet could carry muck that carried seeds, and whether seeds were present in bird droppings. Eventually his observations and calculations persuaded Hooker—always his friend, colleague, and constructive critic. For further references, see Adrian Desmond and James Moore, *Darwin's Sacred Cause* (London: Penguin, 2009). To place Darwin's work on dispersal of species in perspective, see Alan de Queiroz, *The Monkey's Voyage: How Improbable Journeys Shaped the History of Life* (New York: Basic Books, 2014).
3. In the 1820s the Japanese knotweed was rediscovered by a botanist who cultivated it in his nursery in Holland and later marketed it throughout Europe as an ornamental fodder plant. Around the same time, in 1850, the nursery sent a specimen to the Royal Botanical Gardens at Kew. It became highly popular and spread rapidly throughout the United Kingdom through shared cuttings and inappropriate disposal of unwanted plants. By the late nineteenth century it became apparent that the knotweed was destructive to buildings, but by then it was too late to overcome the extreme invasiveness and resilience of this plant.
4. This mass movement of animals, known as the Great Migration, is the largest in the world. Through the course of the year, the animals follow a clockwise, circular migratory pattern as they move in search of greener pastures. This movement culminates in a spectacular fashion toward the middle of the year as millions of wildebeest, zebras, and other herd animals migrate

north from the Serengeti in Tanzania to the Maasai Mara in Kenya. After some months they migrate south again, only to resume the cycle the following year.

5. The Neanderthals lived from about 400,000 years ago until about 30,000 years ago. Stockier and shorter than modern humans, they nonetheless had similar—and perhaps higher—cranial capacity. Archaeological evidence shows they made stone tools, hunted, and used fire. In comparison, little is known about the Denisovans, whose existence was discovered only from the DNA of a finger bone found in a Siberian excavation. Notably, though, genomic studies have revealed interbreeding between Denisovans and Neanderthals, as well as between Neanderthals and *Homo sapiens.* The hominin individuals of *H. floresiensis,* whose skeletons have been retrieved from an Indonesian cave, probably lived 50,000 years ago. This species had a distinctly small stature and small cranial capacity.
6. Alexander Pope, *An Essay on Man* (1733–1734).
7. Ibrahim Abubakar, Robert W. Aldridge, Delan Devakumar, et al., "The UCL-Lancet Commission on Migration and Health: The Health of a World on the Move," *Lancet* 392 (2018): 2606–2654.
8. Unlike migratory birds that make a round trip more than once, each individual monarch butterfly's trip is one-way, with several generations participating in a complete migration cycle. How can they navigate their way across such vast distances without becoming lost? It is thought that the butterflies have special photoreceptors for ultraviolet light in their eyes, and rely on sun position and their antennae circadian clock to maintain their bearing. During their journey, the butterflies accumulate fats, synthesized from sugar in nectar, to help them survive the overwintering period.
9. The red-necked phalarope is one of the rarest birds in the United Kingdom, and until recently little was known about its migration route and wintering grounds. A 2014 study of an adult phalarope tagged in Northern Scotland was the first demonstration of a European breeding bird migrating to the Pacific Ocean. Its total journey was about 60 percent longer than a journey to the Arabian Sea would have been.
10. Niche construction, where organisms actively modify their own and each other's environments, is a fundamental evolutionary process, although its

significance has become increasingly appreciated only in the last two decades or so. There are four major aspects of niche construction theory: (1) Organisms actively modify their environments, introducing a systematic bias to the selective forces generated; (2) exposure of descendants to modified environments (ecological inheritance) affects evolution; (3) acquired characters and by-products systematically influence environments and are therefore evolutionarily significant; and (4) niche construction is a potential evolutionary process underpinning adaptation. See Kevin Laland, Blake Matthews, and Marcus W. Feldman, "An Introduction to Niche Construction Theory," *Evolutionary Ecology* 30 (2016): 191–202.

It is now clear that niche construction concepts, placed within a wider perspective of evolutionary biology and ecology, can greatly facilitate the elucidation of complex biological systems. This is particularly relevant to understanding the evolution of hominin culture and behavior. K. N. Laland, J. Odling-Smee, and M. W. Feldman, "Niche Construction, Biological Evolution, and Cultural Change," *Behavioral and Brain Sciences* 23 (2000): 131–175; K. Laland, *Darwin's Unfinished Symphony: How Culture Made the Human Mind* (Princeton, NJ: Princeton University Press, 2017).

11. Richard Dawkins, *The Extended Phenotype: The Long Reach of the Gene* (Oxford: Oxford University Press, 1999).

12. F. J. Odling-Smee, K. N. Laland, and M. W. Feldman, "Niche Construction," *American Naturalist* 147 (1996): 641–648; F. J. Odling-Smee, K. N. Laland, and M. W. Feldman, *Niche Construction: The Neglected Process of Evolution* (Princeton, NJ: Princeton University Press, 2003).

13. Odling-Smee, Laland, and Feldman term this phenomenon *counteractive niche construction*—we term it *niche modification* because its essential feature, in our view, is that rather than an equilibrium being established between the organism and its constructed niche, humans keep modifying the niche through the use of technology. This uniquely evolved capacity for "niche modification" is essentially the reflection of "cultural evolution," in turn empowered by our cognitive skills and big brains, language abilities that enable social learning and collective knowledge, and our manual dexterity, which allows us to make things (see Chapter 4). Laland et al., *Niche Construction, Biological Evolution, and Cultural Change*, speculate that our modification of

the environment by technology could ultimately lead to our extinction as a species. We hope not. So far, our technological creativity has always gotten us out of trouble. There is no reason to suppose we will not be able to do so in the future, even if our biology and our culture are dramatically altered.

14. Over 40 percent of our genome consists of retrotransposons, which are repetitive DNA elements that can make copies of themselves and insert themselves into new sites. Many of these retrotransposons have been attributed to genome invasion by retroviruses more than twenty-five million years ago. Some retrotransposon insertions have occurred relatively recently in humans and can be used as markers to trace ancestry and study human origins. Although most insertions are neutral, an insertion at a regulatory or protein-coding region of a gene can alter gene expression and give rise to diseases, such as hemophilia, neurofibromatosis, or breast cancer that arises from certain mutations.

15. Mitochondria and chloroplasts evolved from primitive bacterial cells that came to live inside a eukaryotic host cell (a phenomenon known as endosymbiosis; *symbiosis* refers to the interaction between two closely coexisting organisms). The recognition and acceptance of this endosymbiotic theory was mostly spurred by the work of American biologist Lynn Margulis, who also long championed the role of symbiosis as an important evolutionary force.

16. D. E. Elliott and J. V. Weinstock, "Helminth–Host Immunological Interactions: Prevention and Control of Immune-Mediated Diseases," *Annals of the New York Academy of Sciences* 1247 (2012): 83–96.

17. The major hemoglobinopathy that confers a heterozygote advantage against malaria is HbS, found in much of sub-Saharan Africa and parts of the Middle East. Another example involves a different mutation in the same gene; this hemoglobin E variant is prevalent in parts of Southeast Asia, and red blood cells from heterozygous individuals are more resistant to parasite invasion than are cells from homozygotes. Kesinee Chotivanich, Rachanee Udomsangpetch, Kovit Pattanapanyasat, et al., "Hemoglobin E: A Balanced Polymorphism Protective against High Parasitemias and Thus Severe *P Falciparum* Malaria," *Blood* 100 (2002): 1172–1176. Yet another variant, hemoglobin C, protects against malaria, although in this case the homozygous state appears to be more protective.

18. Dawkins, *The Extended Phenotype.*

19. Bacteria and single-cell organisms reproduce by splitting or budding, and thus their genetic information is identical between generations. (They are able to also exchange genetic information by a process called conjugation.) Many other, more complex organisms, including some lizards, can also reproduce without sex via a phenomenon known as parthenogenesis. Some organisms may do this for a number of generations and then switch transiently to sexual reproduction. Sexual reproduction involves gametes—that is, cells with only one copy of each chromosome, such as eggs and sperm in mammals—combining to create a fertilized zygote. The organism then grows with two copies of each gene—one from each parent. Some animals, including many barnacles, which Darwin studied obsessively, are hermaphrodites—that is, they have organs that can produce both eggs and sperm and can self-replicate sexually. There are many theories regarding the reasons sexual reproduction evolved. The most favored theory is that the combining of genetic material from two parents makes the newly formed genotype of the offspring more likely to be resistant to infectious agents that may be in the environment.

20. Alfred Russel Wallace (1823–1913) came from a middle-class family with much more limited resources than Darwin's. Wallace originally trained as a surveyor and was largely self-taught as a naturalist. In 1848 he and another famous explorer-naturalist, Henry Walter Bates (1825–1892), set off for a collecting expedition in the Amazon region. Bates remained in the Amazon for eleven years, and his book *The Naturalist on the River Amazon* (1863) is a classic in travel literature. Wallace returned to Britain in 1852, but many of his specimens had been lost when the ship he was on caught fire and had to be abandoned. Fortunately, he had insurance and could live on the payment until he set off for the Malay Archipelago in 1854. There, he described and collected an enormous number of specimens and made important observations linking geography and species distribution, recognizing that species that were similar to each other in different regions relied on the capacity to have migrated or translocated. The "Wallace line," which distinguishes Southeast Asian biodistribution from Australian–New Guinean distributions, recognizes Wallace's observations of that separation. Wallace became seen as the father of biogeography, and his book *The Geographical Distribution of*

Animals (1876) is the classic volume on the topic. Humboldt and Darwin had also recognized the importance of geography to distributions of flora and fauna. Wallace and Darwin had had some scientific exchanges prior to the events of 1858. Some authors have argued for Wallace's primacy, but Wallace himself acknowledged Darwin's prior contributions and was a staunch defender of Darwin and of natural selection—in 1889 he wrote the book *Darwinism* to rebut a number of critics of the theory. Still, Wallace and Darwin did not agree on everything. In addition to their disagreeing about sexual selection, Darwin strongly disagreed with Wallace's concept that the human mind was the result of spiritual intervention. Despite these disagreements, they remained scientifically collegial, and Wallace walked behind Darwin's coffin at his funeral in Westminster Abbey in 1882. Wallace remained an active naturalist throughout his life, but he also became well known as a spiritualist, which to some extent might have undermined his scientific reputation. He was also very active in socialist causes.

21. The lesser short-tailed bat is one of only two species of bat in New Zealand. It is a "lek" breeder—the males perform to attract females but do not assist in parenting. One of the males' means of attraction is to use their urine as an olfactory attractant. Cory Toth, "Lek Breeding in the Lesser Short-Tailed Bat *(Mystacina tuberculata):* Male Courtship, Female Selection, and the Determinants of Reproductive Strategies," PhD diss., University of Auckland, 2016.

22. A. E. Houde, "Sex Roles, Ornaments and Evolutionary Explanation," *Proceedings of the National Academy of Sciences* 98 (2001): 12857–12859.

23. Ronald Fisher (1894–1962) was perhaps the most distinguished statistician of his era and played a major role in the development of the Modern Synthesis through his work on statistical genetics. In his book *The General Theory of Natural Selection* (1930), Fisher provided a mathematical treatment of how positive feedback between mate preference and sexually determined ornaments could lead to runaway selection. Fisher was a prominent supporter of eugenics.

24. The Irish elk *(Megaloceros giganteus),* despite its popular name, was widespread across Eurasia—the last animals perhaps going extinct in Siberia 8,000 years ago. But its name came from the many skeletons found in Ireland in the eigh-

teenth century, dating to about 11,000 years ago. The elk had the largest antlers known to any deer species and was an important contributor to evolutionary history. When its bones were discovered, it was assumed to represent a yet-to-be-found living species. But about that time the French naturalist and paleontologist Georges Cuvier (1769–1832) recognized and developed the concept of extinctions—namely, that fossils, particularly those of megafauna, were not specimens or relatives of living species that were yet to be discovered but were the bones of extinct species. This was an important contribution to early evolutionary thinking, although Cuvier himself was not a transmutationist and fiercely rejected Lamarck's arguments. There have been various arguments about why the elk became extinct. Some have suggested it may have been a consequence of the handicap of having very large antlers, but perhaps a more plausible explanation is the change in available forage at the end of the Pleistocene epoch. Whether humans played a role in its extinction, as they did for much other megafauna, is unclear.

25. Amotz Zahavi, "Mate Selection: A Selection for a Handicap," *Journal of Theoretical Biology* 53 (1975): 205–214.

26. There is increasing archaeological evidence that ornamentation extended back much earlier in hominin history. It now appears possible that Neanderthals also used some ornamentation, although the significance of these objects as artifacts is not clear. Paul Mellars, "Neanderthal Symbolism and Ornament Manufacture: The Bursting of a Bubble?," *Proceedings of the National Academy of Sciences* 107 (2010): 20147–20148; Dirk L. Hoffmann, Diego E. Angelucci, Valentín Villaverde, et al., "Symbolic Use of Marine Shells and Mineral Pigments by Iberian Neandertals 115,000 Years Ago," *Science Advances* 4 (2018): eaar5255. It appears that by 40,000 years ago there was widespread use of ornament technologies by *Homo sapiens* in Eurasia. Steven L. Kuhn, Mary C. Stiner, David S. Reese, et al., "Ornaments of the Earliest Upper Paleolithic: New Insights from the Levant," *Proceedings of the National Academy of Sciences* 98 (2001): 7641–7646. There is also evidence of earlier developments in the middle Stone Age in Africa. Sally McBrearty and Alison S. Brooks, "The Revolution That Wasn't: A New Interpretation of the Origin of Modern Human Behavior," *Journal of Human Evolution* 39 (2000): 453–563.

27. Evelleen Richards, *Darwin and the Making of Sexual Selection* (Chicago: University of Chicago Press, 2017).

28. Theodore Dobzhansky (1900–1975) was one of the key figures in the Modern Synthesis. He migrated to the United States from Russia in 1927 and worked with the fruit-fly research team led by Thomas Hunt Morgan. In 1937 he wrote *Genetics and the Origin of the Species* (New York: Columbia University Press, 1937), in which he defined evolution as a change in the frequency of an allele within a gene pool—a definition that underpinned the Modern Synthesis. His famous quote comes from the title of a paper he wrote in retirement. T. Dobzhansky, "Nothing in Biology Makes Sense except in the Light of Evolution," *American Biology Teacher* 35 (1973): 125–129. After World War II he translated Ivan Schmalhausen's book on development and evolutionary biology, *Factors of Evolution: The Theory of Stabilizing Selection* (Oxford: Blakiston, 1949)—which, along with Conrad Waddington's work (see Chapter 3), was to underpin the resurgence of appreciation of the role of development in evolution.

Chapter 3 | Inheritance

1. Gregor Mendel (1822–1884) was an experimental biologist as well as an Augustinian friar in a monastery in Brno, now in the Czech Republic. Through his planned breeding experiments, he discovered the basis of particulate rather than blended inheritance, and published his work in an obscure journal. J. G. Mendel, "Versuche über Pflanzenhybriden," *Verhandlungen des naturforschenden Vereines in Brünn* 4 (1865), *Abhandlungen* (1866): 3–47. His observation that inheritance was not simply a blending of parental characteristics and that, instead, depending on the nature of the inheritance, one parent's characteristics could be passed on to the next generation, was in effect the first suggestion that genes are the basis of inheritance. Darwin was not aware of Mendel's work, the significance of which did not become apparent until the early twentieth century, when it was rediscovered by several researchers, including Hugo de Vreis, a Dutch botanist who came up with the word *mutation*. These were the first steps along the path toward modern genetic biology and the Modern Synthesis. The great statistician Ronald Fisher in 1936 claimed, as had others, that Mendel must have adjusted the

numbers in his experiments to make them appear clearer than what the raw data showed. This spawned a vast literature, but historical analyses suggest that Mendel did not fabricate his data. Daniel J. Fairbanks and Bryce Rytting, "Mendelian Controversies: A Botanical and Historical Review," *American Journal of Botany* 88 (2001): 737–752.

2. The leading proponent of the view that selection acts for the good of the species was a British evolutionary biologist, Veru Copner Wynne-Edwards (1906–1997), who wrote an influential book, *Animal Dispersion in Relation to Social Behaviour* (Edinburgh: Oliver and Boyd, 1962), in which he extended the argument for group selection to the concept of species benefit.

3. George C. Williams (1926–2010) vigorously attacked Wynne-Edwards's concept and emphasized the individual as the unit of selection. G. C. Williams, *Adaptation and Natural Selection: A Critique of Some Current Evolutionary Thought* (Princeton, NJ: Princeton University Press, 1966).

4. John Maynard Smith (1920–2004) was a mathematical evolutionary biologist and geneticist who most notably combined game theory with evolutionary biology, considering how individuals would react to each other within a group—either cooperating or cheating. The hawk-dove game is one of the hypothetical games he modeled in his influential book *Evolution and the Theory of Games* (Cambridge: Cambridge University Press, 1982). Smith and Eos Szathmáry, in their book *Evolutionary Transitions* (Oxford: Oxford University Press, 1995), made another major contribution with the concept that evolutionary opportunity arises when there are major transitions, such as when bacteria were first incorporated into cells to become the energy factories as either chloroplasts or mitochondria.

5. Richard Dawkins, *The Selfish Gene* (New York: Oxford University Press, 1976).

6. The Human Genome Project was a publicly funded initiative, led from the United States by Francis Collins but with involvement from other countries, including the United Kingdom, Japan, Germany, and China. The project began in 1990 with the intent of sequencing the three billion nucleotides that make up the human genome. At the same time, through his company Celera Genomics, Craig Venter attempted to do the same, funded by investors. Both projects culminated in initial sequencing drafts in 2001, with

more-complete data produced in subsequent years. The data indicated that the number of coding genes in the human genome is about 22,000, far fewer than previously thought. It also showed that there is a high homology of coding genes in very different species (highlighting the unity of life, as Darwin had first proposed) and that most of the DNA does not code for proteins. Although the estimates vary according to the methods used, humans have about 60 percent genetic homology to fruit flies in coding regions and 23 percent to baker's yeast. Mathematical approaches that take into account estimated mutation rates enable researchers to establish the approximate point at which two organisms shared a last common ancestor. For example, the last common ancestor between gorillas and humans likely lived in Africa about ten million years ago. Aylwyn Scally, Julien Y. Dutheil, LaDeana W. Hillier, et al., "Insights into Hominid Evolution from the Gorilla Genome Sequence," *Nature* 483 (2012): 169–175.

7. The debate about the levels of selection, which has continued for decades, has relied on various models and issues regarding the evolution of altruism. The concept of kin selection advanced by Hamilton posited that altruism could be explained as promoting inclusive fitness by promoting the survival of genes shared with kin—the argument being that in the evolutionary past, members of a group likely had kin relationships. W. D. Hamilton, "The Genetical Evolution of Social Behaviour," *Journal of Theoretical Biology* 7 (1964): 1–16. There are echoes here of the selfish gene and gene-level selection. The alternative model of reciprocity, based on game theory (that altruism arises because a favor done now is likely to be reciprocated later, so both parties gain advantage), is also based on the individual as the target of selection. But as Darwin implied, social norms that develop within a group can be the basis of selection. This view is exemplified in the work of Robert Boyd and Peter Richerson: Boyd and Richerson, *Not by Genes Alone: How Culture Transformed Human Evolution* (Chicago: University of Chicago Press, 2006); Boyd and Richerson, *Culture and the Evolutionary Process* (Chicago: University of Chicago Press, 1988); Boyd, *A Different Kind of Animal: How Culture Transformed Our Species* (Princeton, NJ: Princeton University Press, 2017). Indeed, it is difficult to see how some aspects of cultural evolution could be explained without reference to group-level selection. Some genetic determinists still reject the concept of

multilevel selection, convinced that Dawkins and Williams have refuted that possibility. But as David Sloan Wilson has elucidated, multilevel selection is fully compatible with genetic as well as cultural evolution. David Sloan Wilson, "Reaching a New Plateau for the Acceptance of Multilevel Selection," Evolution Institute (2017), https://evolution-institute.org/focus-article/reaching-a-new-plateau-for-the-acceptance-of-multilevel-selection/. So the issue may have been more about some of the simplifying assumptions made in the early models. Indeed, Hamilton later acknowledged the potential for group selection. For more extensive discussions, see David Sloan Wilson, *Does Altruism Exist? Culture, Genes, and the Welfare of Others* (New Haven, CT: Yale University Press, 2015); D. S. Wilson and E. O. Wilson, "Rethinking the Theoretical Basis of Sociobiology," *Quarterly Review of Biology* 82 (2007): 327–348; Samir Okasha, *Evolution and the Levels of Selection* (Oxford: Oxford University Press, 2006); Joseph Heinrich, *The Secret of Our Success: How Culture Is Driving Human Evolution, Domesticating Our Species and Making Us Smarter* (Princeton, NJ: Princeton University Press, 2015).

8. Haplodiploidy is a form of sex determination that occurs in bees, wasps, and ants. Males have only one set of chromosomes (haploid), and females have two sets (diploid). The males develop from unfertilized eggs of the queen. Females result from sexual reproduction. Because of this particular reproductive strategy, if a queen mates only with a single drone (a male) that is one of her offspring, then any two daughters will share three-fourths of their genome, making them more closely related than two sisters in fully diploid species. This has been suggested as one reason for the intense sociality of these hive species.
9. For interesting discussions on the differing concepts of the gene, see Lenny Moss, *What Genes Can't Do* (Cambridge, MA: MIT Press, 2004); and Evelyn Fox Keller, *The Century of the Gene* (Cambridge, MA: Harvard University Press, 2002).
10. F. H. Crick, "On Protein Synthesis," *Symposia of the Society for Experimental Biology* 12 (1958): 138–163.
11. But simple metaphors such as "blueprint" have a persistent life in science, especially popular science, and continue to be used despite their inaccuracy.

For instance, see Robert Plomin, *Blueprint: How DNA Makes Us Who We Are* (Cambridge, MA: MIT Press, 2018).

12. For a discussion of the misconceptions about heredity and the data from personal genomes, see Adam Rutherford, *A Brief History of Everyone Who Ever Lived: The Stories in Our Genes* (London: Orion, 2016).

13. Darwin never excluded the role of use-inheritance in his thinking, perhaps because of the lack of a realistic model of inheritance. He realized that beyond hard inheritance, environmental factors in some way might influence the phenotype, hence his comment in *The Origin of Species:* "It is well known to furriers that animals of the same species have thicker and better fur the further north they live; but who can tell how much of this difference may be due to the warmest-clad individuals having been favored and preserved during many generations, and how much to the action of the severe climate?"

14. August Weismann (1834–1914), a German biologist, was a critical figure in evolutionary thought and genetics. In 1868 he published, in German, *On the Justification of Darwinian Theory.* He was a fervent supporter of Darwin and very opposed to Lamarckian concepts—hence his mouse experiment, flawed as he knew it to be. His major contribution, which became known as Weismannism, was the theory that germ cells (that is, sperm and eggs) are immune to influences from the environment and, thus, that genetic inheritance is not influenced by the environment. This was published in his book *The Germ Plasm: A Theory of Heredity* in 1892 (translated into English a year later). This view was accepted as absolute dogma through most of the twentieth century, and was part of reason arguments such as Conrad Waddington's were excluded in the discussions that formulated the Modern Synthesis. We now know that the so-called "Weismann barrier" is not absolute. Epigenetic observations show that sperm and ova can be influenced by external factors, and microRNAs have been shown to cross the somatic-gonadal barrier.

15. For a review of Sonneborn's work, see J. R. Preer, "Sonneborn and the Cytoplasm," *Genetics* 172 (2006): 1373–1377.

16. Kammerer's book *Environmental Vitalism: The Inheritance of Acquired Characteristics* was republished with an introduction by John Gist (Gold Beach, OR: High Sierra, 2003).

17. William Bateson (1861–1926) was a British biologist who coined the term *genetics*. He was a major contributor to the early discussions reconciling Darwinian concepts with particulate inheritance. That debate was resolved in the decade after his death in what became known as the Modern Synthesis. Like Darwin, he was very interested in the phenomena of phenotypic variation, which he summarized in his book *Materials for the Study of Variation* (1894). He was a cousin, two generations removed, of the authors' late colleague Sir Patrick Bateson FRS.
18. Trofim Lysenko (1898–1967) was a Soviet agronomist and pseudoscientist. Lysenko claimed to have rediscovered vernalization—a method already in use to treat wheat seeds with moisture and cold in the hope of increasing spring yields. On the basis of minimum evidence he made great claims and became much favored by Josef Stalin. Lysenko strongly rejected all concepts of genetics and of the hereditary role of DNA. He had his own version of environmental inheritance, which he claimed had no linkage to Lamarckism, although there were some parallels. As he rose to be all-powerful in Soviet biology, he ensured that his geneticist opponents were persecuted. Some, such as Nikolai Vavilov, a prominent geneticist and a vocal critic of Lysenko, died in a gulag; others emigrated or ceased their work. Lysenko fell into disfavor after Stalin's death, but Soviet Russian biology had been set back decades, and some very important science—especially in the evolutionary sciences, such as the work of Ivan Schmalhausen (a pioneering thinker in developmental biology)—had effectively been suppressed or lost. See Loren Graham, *Lysenko's Ghost: Epigenetics and Russia* (Cambridge, MA: Harvard University Press, 2016).
19. A. O. Vargas, "Did Paul Kammerer Discover Epigenetic Inheritance? A Modern Look at the Controversial Midwife Toad Experiments," *Journal of Experimental Zoology Part B: Molecular and Developmental Evolution* 312B (2009): 667–678; Alexander O. Vargas, Quirin Krabichler, and Carlos Guerrero-Bosagna, "An Epigenetic Perspective on the Midwife Toad Experiments of Paul Kammerer (1880–1926)," *Journal of Experimental Zoology Part B: Molecular and Developmental Evolution* 328 (2017): 179–192.
20. Conrad Waddington (1905–1975) was a British developmental and evolutionary biologist, who along with the brilliant Soviet scientist Ivan Schmalhausen (1880–1963) had argued for a place for developmental biology within the Darwinian synthesis. Both believed that environmental influences acting

on the developing organism could influence evolutionary processes. Waddington proposed the concept of genetic assimilation, by which environmentally induced phenotypic change could over time be incorporated into the genome. This remains a speculative but important concept in "evo-devo" and in understanding the relationship between developmental plasticity and evolutionary processes. Mary Jane West-Eberhard, in her magnum opus *Developmental Plasticity and Evolution* (Oxford: Oxford University Press, 2003), discusses this phenomenon at length. It has been further examined by Patrick Bateson and Peter Gluckman in *Plasticity, Robustness, Development and Evolution* (Cambridge: Cambridge University Press, 2011).

Waddington also introduced the concept of canalization and the term *epigenetics*. His book *The Strategy of the Genes* (London: George Allen and Unwin, 1957) is a classic in the evolution of broader concepts of post neo-Darwinian thinking—the concept of an extended evolutionary synthesis that incorporates development and the possibilities of non-Mendelian inheritance. Waddington illustrated his concept of canalization and developmental plasticity by means of his epigenetic landscape diagram in his book *Organisers and Genes* (Cambridge: Cambridge University Press, 1940). He then added to it by illustrating the underside of the landscape. For a detailed description of the significance of these figures, see Adam R. Navis, "Epigenetic Landscape," Embryo Project Encyclopedia, October 30, 2007, https://embryo.asu.edu/pages/epigenetic-landscape. Waddington was a well-known Marxist sympathizer and a socially active writer. He also attended the Hunstanton Mill meetings of the Theoretical Biology Club, where he would have met Karl Popper and others.

Concepts of landscapes had already been applied in evolutionary biology, most notably by Sewell Wright (1889–1988), one of the architects of the Modern Synthesis. He had used the landscape analogy to highlight the processes of speciation and the concepts of natural selection and genetic drift. Natural selection occurred by gradual ascending of a slope (the height reflecting relative fitness), and species were separated into different niches represented by different peaks. Richard Dawkins used the same landscape concepts, and the idea that peaks in the landscape were points of evolution, in his popular science book *Climbing Mount Improbable* (New York: Norton, 1996).

21. See Navis, "Epigenetic Landscape."

22. More recently, royal jelly has been shown to contain some of the small noncoding RNAs which induce the epigenetic changes that set the developmental fate of the larvae. Kegan Zhu, Minghui Liu, Zheng Fu, et al., "Plant microRNAs in Larval Food Regulate Honeybee Caste Development," *PLOS Genetics* 13 (2017): e1006946.

23. Bhavatharini Kasinathan, Kami Ahmad, and Harmit S. Malik, "Waddington Redux: *De Novo* Mutations Underlie the Genetic Assimilation of Stress-Induced Phenocopies in *Drosophila melanogaster*," *Genetics* 207 (2017): 49–51.

24. R. Holliday and J. E. Pugh, "DNA Modification Mechanisms and Gene Activity during Development," *Science* 187 (1975): 226–232.

25. See Navis, "Epigenetic Landscape."

26. There is now vast empirical and theoretical literature on non-genomic inheritance and the role of epigenetic processes. Some of the most comprehensive reviews are Eva Jablonka and Gal Raz, "Transgenerational Epigenetic Inheritance: Prevalence, Mechanisms, and Implications for the Study of Heredity and Evolution," *Quarterly Review of Biology* 84 (2009): 131–176; Russell Bonduriansky and Troy Day, *Extended Heredity: A New Understanding of Inheritance and Evolution* (Princeton, NJ: Princeton University Press, 2018); and Sabine Schaefer and Joseph H. Nadeau, "The Genetics of Epigenetic Inheritance: Modes, Molecules, and Mechanisms," *Quarterly Review of Biology* 90 (2015): 381–415. For a discussion of the multiple possible mechanisms of non-genomic inheritance, see P. D. Gluckman, M. A. Hanson, and A. S. Beedle, "Non-Genomic Transgenerational Inheritance of Disease Risk," *Bioessays* 29 (2007): 145–154.

27. David Barker (1938–2013) was physician and epidemiologist at the University of Southampton, England. His observations on the relationship between birth weight and later disease risk laid the foundation for the field of study now known as the Developmental Origins of Health and Disease (DOHaD). Barker's team produced a plethora of studies linking the conditions of fetal and infant life to later risks of noncommunicable disease, and his work and collegiality spawned a large group of studies, including those of the authors.

28. Keith M. Godfrey, Allan Sheppard, Peter D. Gluckman, et al., "Epigenetic Gene Promoter Methylation at Birth Is Associated with Child's Later Adiposity," *Diabetes* 60 (2011): 1528–1534.

29. For example, Hong Pan, Xinyi Lin, Yonghui Wu, et al., "*HIF3A* Association with Adiposity: The Story Begins before Birth," *Epigenomics* 7 (2015): 937–950; Karen Lillycrop, Robert Murray, Clara Cheong, et al., "Anril Promoter DNA Methylation: A Perinatal Marker for Later Adiposity," *EBioMedicine* 19 (2017): 60–72.
30. Prasoon Agarwal, Taylor S. Morriseau, Stephanie M. Kereliuk, et al., "Maternal Obesity, Diabetes during Pregnancy and Epigenetic Mechanisms That Influence the Developmental Origins of Cardiometabolic Disease in the Offspring," *Critical Reviews in Clinical Laboratory Sciences* 55 (2018): 71–101; Priyanka Parmar, Estelle Lowry, Giovanni Cugliari, et al., "Association of Maternal Prenatal Smoking *GFI1*-Locus and Cardio-Metabolic Phenotypes in 18,212 Adults," *EBioMedicine* 38 (2018): 206–216. Recent evidence in mice suggests that a maternal high-fat diet may even have cognitive impacts over several generations. Gitalee Sarker and Daria Peleg-Raibstein, "Maternal Overnutrition Induces Long-Term Cognitive Deficits across Several Generations," *Nutrients* 11 (2018): 7.
31. Benjamin R. Carone, Lucas Fauquier, Naomi Habib, et al., "Paternally Induced Transgenerational Environmental Reprogramming of Metabolic Gene Expression in Mammals," *Cell* 143 (2010): 1084–1096; T. Fullston, E. M. C. Ohlsson Teague, N. O. Palmer, et al., "Paternal Obesity Initiates Metabolic Disturbances in Two Generations of Mice with Incomplete Penetrance to the F_2 Generation and Alters the Transcriptional Profile of Testis and Sperm microRNA Content," *FASEB Journal* 27 (2013): 4226–4243.
32. Valérie Grandjean, Sandra Fourré, Diana Andrea Fernandes De Abreu, et al., "RNA-Mediated Paternal Heredity of Diet-Induced Obesity and Metabolic Disorders," *Scientific Reports* 5 (2015): 18193; K. D. Wagner, N. Wagner, H. Ghanbarian, et al., "RNA Induction and Inheritance of Epigenetic Cardiac Hypertrophy in the Mouse," *Developmental Cell* 14 (2008): 962–969.
33. Sam Houfflyn, Christophe Matthys, and Adelheid Soubry, "Male Obesity: Epigenetic Origin and Effects in Sperm and Offspring," *Current Molecular Biology Reports* 3 (2017): 288–296; Emma L. Marczylo, Akwasi A. Amoako, Justin C. Konje, et al., "Smoking Induces Differential miRNA Expression in Human Spermatozoa: A Potential Transgenerational Epigenetic Concern?," *Epigenetics* 7 (2012): 432–439.

34. Carlos Guerrero-Bosagna, Trevor R. Covert, M. M. Haque, et al., "Epigenetic Transgenerational Inheritance of Vinclozolin Induced Mouse Adult Onset Disease and Associated Sperm Epigenome Biomarkers," *Reproductive Toxicology* 34 (2012): 694–707.

35. Wenfei Sun, Hua Dong, Anton S. Becker, et al., "Cold-Induced Epigenetic Programming of the Sperm Enhances Brown Adipose Tissue Activity in the Offspring," *Nature Medicine* 24 (2018): 1372–1383.

36. The Dutch Hunger Winter, or *Hongerwinter,* was an intentionally induced famine imposed by Nazi occupiers of the Netherlands in September 1944 in response to Dutch resistance activities. It became progressively more severe until the Allies displaced the occupation in May 1945. In some places food intake fell from more than 2,000 calories per day to about 600 calories per day. This is an unusual famine in that the timing of its onset and end are well documented. A number of studies have been made on women who experienced the famine while pregnant and on their offspring and grandoffspring. Studies of those who were pregnant before or after the famine, or only partially during the famine, have produced a number of important observations from this ugly episode of war. The first reported observations were that the offspring were more likely to be obese as adults, but only if the mother had been undernourished in the first half of pregnancy. G. P. Ravelli, Z. A. Stein, and M. W. Susser, "Obesity in Young Men after Famine Exposure in Utero and Early Infancy," *New England Journal of Medicine* 295 (1976): 349–353. Many other studies by a number of investigators, including from Barker's group, have shown that offspring subject to famine have greater risks of noncommunicable disease, in line with the DOHaD paradigm. Tessa J. Roseboom, Jan H. P. van der Meulen, Anita C. J. Ravelli, et al., "Effects of Prenatal Exposure to the Dutch Famine on Adult Disease in Later Life: An Overview," *Molecular and Cellular Endocrinology* 185 (2001): 93–98. More-recent observations show epigenetic effects on the offspring. B. T. Heijmans, E. W. Tobi, A. D. Stein, et al., "Persistent Epigenetic Differences Associated with Prenatal Exposure to Famine in Humans," *Proceedings of the National Academy of Sciences* 105 (2008): 17046–17049; Elmar W. Tobi, Roderick C. Slieker, René Luijk, et al., "DNA Methylation as a Mediator of the Association between Prenatal Adversity and Risk Factors for Metabolic Disease in Adulthood," *Science Advances* 4 (2018): eaao4364; M. V. Veenendaal,

P. M. Costello, K. A. Lillycrop, et al., "Prenatal Famine Exposure, Health in Later Life and Promoter Methylation of Four Candidate Genes," *Journal of Developmental Origins of Health and Disease* 3 (2012): 450–457. Male-line mediated phenotypic effects in grandoffspring have also been found. M. V. Veenendaal, R. C. Painter, S. R. de Rooij, et al., "Transgenerational Effects of Prenatal Exposure to the 1944–1945 Dutch Famine," *BJOG: An International Journal of Obstetrics & Gynaecology* 120 (2013): 548–554.

37. A limited study of historical records suggests that prepubertal experience of famine by paternal grandmothers had adverse impacts on their granddaughters' risks of cardiovascular mortality. Lars Bygren, Petter Tinghog, John Carstensen, et al., "Change in Paternal Grandmothers' Early Food Supply Influenced Cardiovascular Mortality of the Female Grandchildren," *BMC Genetics* 15 (2014): 12. Other studies of the offspring of paternal smokers showed an association with offspring obesity. M. E. Pembrey, L. O. Bygren, G. Kaati, et al., "Sex-Specific, Male-Line Transgenerational Responses in Humans," *European Journal of Human Genetics* 14 (2006): 159–166. Although these studies are suggestive, there are alternate explanations for these data, and epigenetic inheritance is not proven by these studies.

38. Adelheid Soubry, Cathrine Hoyo, Randy L. Jirtle, et al., "A Paternal Environmental Legacy: Evidence for Epigenetic Inheritance through the Male Germ Line," *Bioessays* 36 (2014): 359–371; Adelheid Soubry, Joellen Schildkraut, Amy Murtha, et al., "Paternal Obesity Is Associated with IGF2 Hypomethylation in Newborns: Results from a Newborn Epigenetics Study (NEST) Cohort," *BMC Medicine* 11 (2013): 29; A. Soubry, S. K. Murphy, F. Wang, et al., "Newborns of Obese Parents Have Altered DNA Methylation Patterns at Imprinted Genes," *International Journal of Obesity* 39 (2015): 650–657; Ida Donkin, Soetkin Versteyhe, Lars R. Ingerslev, et al., "Obesity and Bariatric Surgery Drive Epigenetic Variation of Spermatozoa in Humans," *Cell Metabolism* 23 (2016): 369–378. For a more extensive review of the underlying molecular mechanisms, see Louise Ruby Høj Illum, Stine Thorhauge Bak, Sten Lund, et al., "DNA Methylation in Epigenetic Inheritance of Metabolic Diseases through the Male Germ Line," *Journal of Molecular Endocrinology* 60 (2018): R39–R56.

39. Anna Fogel, Lisa R. Fries, Keri McCrickerd, et al., "Prospective Associations between Parental Feeding Practices and Children's Oral Processing Be-

haviours," *Maternal & Child Nutrition* 15 (2019): e12635; Anna Fogel, Lisa R. Fries, Keri McCrickerd, et al., "Oral Processing Behaviours That Promote Children's Energy Intake Are Associated with Parent-Reported Appetitive Traits: Results from the GUSTO Cohort," *Appetite* 126 (2018): 8–15.

40. Denis Noble, *Dance to the Tune of Life: Biological Relativity* (Cambridge: Cambridge University Press, 2016).

41. This is modeled diagrammatically in P. Gluckman and M. Hanson, *The Fetal Matrix: Evolution, Development, and Disease* (Cambridge: Cambridge University Press, 2005).

42. John Waterlow (1919–2010) was a doyen in the study of child malnutrition. He spent much of his early research career in Jamaica, where in 1954 he established the Tropical Metabolism Research Unit, a major research center for malnutrition where world-class research still continues.

43. David Barker published numerous groundbreaking papers, such as: D. J. Barker and C. Osmond, "Infant Mortality, Childhood Nutrition, and Ischaemic Heart Disease in England and Wales," *Lancet* 1 (1986): 1077–1081; D. J. P. Barker, P. D. Winter, C. Osmond, et al., "Weight in Infancy and Death from Ischaemic Heart Disease," *Lancet* 2 (1989): 577–580; D. J. P. Barker, A. R. Bull, C. Osmond, and S. J. Simmonds, "Fetal and Placental Size and Risk of Hypertension in Adult Life," *British Medical Journal* 301 (1990): 259–262; C. N. Hales, D. J. P. Barker, P. M. S. Clark, et al., "Fetal and Infant Growth and Impaired Glucose Tolerance at Age 64," *British Medical Journal* 303 (1991): 1019–1022.

    Others at about the same time had made similar observations on the longer-term consequences of a poor start to life. For instance: A. Forsdahl, "Living Conditions in Childhood and Subsequent Development of Risk Factors for Arteriosclerotic Heart Disease: The Cardiovascular Survey in Finnmark, 1974–1975," *Journal of Epidemiology and Community Health* 32 (1978): 34–37; M. W. Higgins, J. B. Keller, H. L. Metzner, et al., "Studies of Blood Pressure in Tecumseh, Michigan: II. Antecedents in Childhood of High Blood Pressure in Young Adults," *Hypertension* 2 (1980): 117–123; V. Notkola, S. Punsar, M. J. Karvonen, and J. Haapakoski, "Socioeconomic Conditions in Childhood and Mortality and Morbidity Caused by Coronary Heart Disease in Adulthood in Rural Finland," *Social Science & Medicine* 21

(1985): 517–523; M. E. Wadsworth, H. A. Cripps, R. E. Midwinter, and J. R. Colley, "Blood Pressure in a National Birth Cohort at the Age of 36 Related to Social and Familial Factors, Smoking, and Body Mass," *British Medical Journal (Clinical Research Ed.)* 291 (1985): 1534–1538. But it was Barker who had recognized the broader significance of these studies.

44. C. N. Hales and D. J. Barker, "Type 2 (Non-Insulin-Dependent) Diabetes Mellitus: The Thrifty Phenotype Hypothesis," *Diabetologia* 35 (1992): 595–601.

45. Gluckman and Hanson, *The Fetal Matrix*; C. W. Kuzawa, P. D. Gluckman, and M. A. Hanson, "Developmental Perspectives on the Origin of Obesity, in *Adipose Tissue and Adipokines in Health and Disease,* ed. Giamila Fantuzzi and Theodore Mazzone, 207–219 (Totowa, NJ: Humana Press, 2007); Mark A. Hanson and P. D. Gluckman, "Early Developmental Conditioning of Later Health and Disease: Physiology or Pathophysiology?," *Physiological Reviews* 94 (2014): 1027–1076.

46. P. D. Gluckman and M. A. Hanson, "Living with the Past: Evolution, Development, and Patterns of Disease," *Science* 305 (2004): 1733–1736; Peter D. Gluckman, Mark Hanson, Paul Zimmet, et al., "Losing the War against Obesity: The Need for a Developmental Perspective," *Science Translational Medicine* 3 (2011): 93cm19.

47. See T. M. Lee and I. Zucker, "Vole Infant Development Is Influenced Perinatally by Maternal Photoperiodic History," *American Journal of Physiology* 255 (1988): R831–R838. More recently, such predictive adaptive responses have been observed in cuttlefish embryos, which decrease their ventilation rate as an adaptive response to the presence of predators. Nawel Mezrai, Lorenzo Arduini, Ludovic Dickel, et al., "Awareness of Danger Inside the Egg? Evidence of Innate and Learned Predator Recognition in Cuttlefish Embryo," *bioRxiv* (2018), https://doi.org/10.1101/508853.

48. Meir Paul Pener and Stephen J. Simpson, "Locust Phase Polyphenism: An Update," *Advances in Insect Physiology* 36 (2009): 1–272.

49. Ivan Gomez-Mestre, Saurabh Kulkarni, and Daniel R. Buchholz, "Mechanisms and Consequences of Developmental Acceleration in Tadpoles Responding to Pond Drying," *PLOS One* 8 (2013): e84266.

50. P. D. Gluckman, M. A. Hanson, and H. G. Spencer, "Predictive Adaptive Responses and Human Evolution," *Trends in Ecology & Evolution* 20 (2005):

527–533; Patrick Bateson, Peter Gluckman, and Mark Hanson, "The Biology of Developmental Plasticity and the Predictive Adaptive Response Hypothesis," *Journal of Physiology* 592 (2014): 2357–2368.

51. P. Bateson, "Fetal Experience and Good Adult Design," *International Journal of Epidemiology* 30 (2001): 928–934.

52. T. E. Forrester, A. V. Badaloo, M. S. Boyne, et al., "Prenatal Factors Contribute to Emergence of Kwashiorkor or Marasmus in Response to Severe Undernutrition: Evidence for the Predictive Adaptation Model," *PLOS One* 7 (2012): e35907; Allan Sheppard, Sherry Ngo, Xiaoling Li, et al., "Molecular Evidence for Differential Long-Term Outcomes of Early Life Severe Acute Malnutrition," *EBioMedicine* 18 (2017): 274–280.

53. M. J. Meaney, "Maternal Care, Gene Expression, and the Transmission of Individual Differences in Stress Reactivity across Generations," *Annual Review of Neuroscience* 24 (2001): 1161–1192.

54. I. C. G. Weaver, N. Cervoni, F. A. Champagne, et al., "Epigenetic Programming by Maternal Behavior," *Nature Neuroscience* 7 (2004): 847–854.

55. Another example is the red squirrel, in which offspring of mothers exposed to higher population density cues via recorded calls were found to grow faster—an effect mediated by maternal steroid levels. Ben Dantzer, Amy E. M. Newman, Rudy Boonstra, et al., "Density Triggers Maternal Hormones That Increase Adaptive Offspring Growth in a Wild Mammal," *Science* 340 (2013): 1215–1217.

56. A. Qiu, T. T. Anh, Y. Li, et al., "Prenatal Maternal Depression Alters Amygdala Functional Connectivity in 6-Month-Old Infants," *Translational Psychiatry* 5 (2015): e508.

57. On the question of whether evolutionary theory needs to be reframed, some useful references are K. Laland, T. Uller, M. Feldman, et al., "Does Evolutionary Theory Need a Rethink?," *Nature* 514 (2014): 161–164; Ueli Grossniklaus, William G. Kelly, Bill Kelly, et al., "Transgenerational Epigenetic Inheritance: How Important Is It?," *Nature Reviews Genetics* 14 (2013): 228–235; K. N. Laland, T. Uller, M. Feldman, et al., "The Extended Evolutionary Synthesis: Its Structure, Assumptions and Predictions," *Proceedings of the Royal Society B: Biological Sciences* 282 (2015): 1–14.

Chapter 4 | Culture

1. Although today we would have difficulty with the language used in his time to discuss this issue, Darwin came from a family that believed in the abolition of slavery. He repeatedly debated this issue with FitzRoy, the captain of HMS *Beagle.* Adrian Desmond and James Moore, *Darwin's Sacred Cause* (London: Penguin, 2009).
2. Sir Richard Burton (1821–1890) was an early social anthropologist and a noted explorer of East Africa, South America, and the Arabian Peninsula. His interest in cultural diversity extended to matters that were somewhat shocking in more puritanical Victorian circles. He described detailed sexual practices and measured the lengths of penises of residents in the areas he visited. He also translated much erotic literature, including *The Book of the Thousand Nights and a Night* (1885) (popularly known as *The Arabian Nights*).
3. The discovery and dating of stone tools and fossils in archaeological sites has shown that early humans were capable of tool making. Archaeologists have identified scrapers, choppers, and cutters that are simple in design but would have required substantial cognitive capacity and manual dexterity to plan and shape them. More recent discoveries of older specimens now suggest that tool making began even earlier, by more ancient species such as the Australopithecines.
4. Richard O. Prum, *The Evolution of Beauty: How Darwin's Forgotten Theory of Mate Choice Shapes the Animal World—and Us* (New York: Doubleday, 2017).
5. M. Nadal, M. Capó, Enric Munar, et al., "Constraining Hypotheses on the Evolution of Art and Aesthetic Appreciation," in *Neuroaesthetics,* ed. Martin Shov and Oshin Vartanian, 103–129 (Abingdon, UK: Routledge, 2008).
6. The contention stems from the reliance on indirect methods to study language in extinct hominins. The soft tissues of the larynx do not fossilize, therefore the methods for studying the ability of various hominins to vocalize (or, indeed, to verbally interact) have to be inferential, and based mostly on knowledge of brain anatomy, lithic technology, and social structure and behavior. As a result, estimates for when language originated range, wildly, from 400,000 years ago to 50,000 years ago.

7. There is an extensive literature on the evolution of language. More relevant to the context of this book is the relationship of language to human cultural evolution. For extensive discussions of this relationship, see M. Pagel, *Wired for Culture: The Natural History of Human Cooperation* (London: Allen Lane, 2012); Joseph Heinrich, *The Secret of Our Success: How Culture Is Driving Human Evolution, Domesticating Our Species and Making Us Smarter* (Princeton, NJ: Princeton University Press, 2015); and K. Laland, *Darwin's Unfinished Symphony: How Culture Made the Human Mind* (Princeton, NJ: Princeton University Press, 2017).

8. Jeremy Button was the name given to a Fuegian of the Yaghan people who had been abducted, along with three other young Fuegians, in Tierra del Fuego by Captain FitzRoy on an earlier expedition of HMS *Beagle* in 1830. The four were taken hostage after one of FitzRoy's boats was stolen. They were taken back to England to be "civilized" and Christianized, with the intent that they be returned to Tierra del Fuego to assist in developing a mission. Of the four, one, who was called Boat Memory, died soon after arriving in England. Button, along with the other two survivors, called York Minster and Fuegia Basket, was returned to Tierra del Fuego a year later on the same voyage as Darwin on the *Beagle.* Button soon shed his European clothes and habits, and when the *Beagle,* with Darwin on board, made another visit only a few months later, he was seen living as an indigenous person, naked except for a loincloth. Some years later, in 1859, Button was alleged to have taken part in a massacre of missionaries by the Yaghan. Button, who had retained his abilities in English, denied involvement.

9. See Claes Andersson and Dwight Read, "The Evolution of Cultural Complexity: Not by the Treadmill Alone," *Current Anthropology* 57 (2016): 261–286.

10. S. Kawamura, "The Process of Sub-culture Propagation among Japanese Macaques," *Primates* 2 (1959): 43–60; T. Matsuzawa, "Sweet-Potato Washing Revisited: 50th Anniversary of the *Primates* Article," *Primates* 56 (2015): 285–287.

11. Simon E. Fisher, "Evolution of Language: Lessons from the Genome," *Psychonomic Bulletin & Review* 24 (2017): 34–40; Dieter G. Hillert, "On the Evolving Biology of Language," *Frontiers in Psychology* 6 (2015): 1796.

12. A valuable review of these different approaches to understanding human behavior through an evolutionary lens is K. N. Laland and G. R. Brown, *Sense and Nonsense: Evolutionary Perspectives on Human Behaviour* (Oxford: Oxford University Press, 2002).

13. Leda Cosmides and John Tooby, "Evolutionary Psychology: New Perspectives on Cognition and Motivation," *Annual Review of Psychology* 64 (2013): 201–229.

14. Steven Pinker, *How the Mind Works* (New York: Norton, 1999); Robert Wright, *The Moral Animal: Why We Are the Way We Are: The New Science of Evolutionary Psychology* (New York: Vintage, 1995).

15. Elliott Sober, "The Evolution of Altruism: Correlation, Cost, and Benefit," *Biology and Philosophy* 7 (1992): 177–187; E. Sober and D. S. Wilson, *Unto Others: The Evolution and Psychology of Unselfish Behavior* (Cambridge, MA: Harvard University Press, 1998); David Sloan Wilson, *Does Altruism Exist? Culture, Genes, and the Welfare of Others* (New Haven, CT: Yale University Press, 2015).

16. See John Dupré, "Against Maladaptationism: Or, What's Wrong with Evolutionary Psychology?," in *Processes of Life: Essays in the Philosophy of Biology*, ed. John Dupré (Oxford: Oxford Scholarship Online, 2012).

17. Robert Foley, "The Adaptive Legacy of Human Evolution: A Search for the Environment of Evolutionary Adaptedness," *Evolutionary Anthropology* 4 (1995): 194–203.

18. Noam Chomsky, *Rules and Representations* (New York: Columbia University Press, 1980).

19. René Dubos, *Man Adapting* (New Haven, CT: Yale University Press, 1965).

20. E. A. Maguire, D. G. Gadian, I. S. Johnsrude, et al., "Navigation-Related Structural Change in the Hippocampi of Taxi Drivers," *Proceedings of the National Academy of Sciences of the United States of America* 97 (2000): 4398–4403.

21. Kevin C. Bickart, Christopher I. Wright, Rebecca J. Dautoff, et al., "Amygdala Volume and Social Network Size in Humans," *Nature Neuroscience* 4 (2011): 163–164; R. Kanai, B. Bahrami, R. Roylance, and G. Rees, "Online Social Network Size Is Reflected in Human Brain Structure," *Proceedings of the Royal Society B: Biological Sciences* 279 (2012): 1327–1334.

22. A large brain-imaging study of ten-year-olds has found that excessive screen time from television / video viewing, video game playing, and social media use was linked to subtle alterations in brain structure, with possible functional consequences. Martin P. Paulus, Lindsay M. Squeglia, Kara Bagot, et al., "Screen Media Activity and Brain Structure in Youth: Evidence for Diverse Structural Correlation Networks from the ABCD Study," *NeuroImage* 185 (2019): 140–153. A longitudinal study has also found that greater exposure to screen time at the age of two years was correlated with obesity, illnesses, poor motor skills, and hyperactivity at age 4½ years. Tom Stewart, Scott Duncan, Caroline Walker, et al., *Effects of Screen Time on Preschool Health and Development* (Wellington, NZ: Ministry of Social Development, 2019). Furthermore, it has been recognized that in the context of today's media-rich society, exposure to emotion-arousing media may have particular impact on adolescents, who have greater sensitivity to peer acceptance and rejection but have not yet fully matured in brain regions that modulate executive function and impulse control. Eveline A. Crone and Elly A. Konijn, "Media Use and Brain Development during Adolescence," *Nature Communications* 9 (2018): 588. For further discussion, see Yolanda Reid Chassiakos, Jenny Radesky, Dimitri Christakis, et al., "Children and Adolescents and Digital Media," *Pediatrics* 138 (2016): e20162593. Still, it is a challenge to fully understand the functional consequences of screen time and digital technologies. Amy Orben and Andrew K. Przybylski, "The Association between Adolescent Well-Being and Digital Technology Use," *Nature Human Behaviour* (2019), https://doi.org/10.1038/s41562-018-0506-1; Vaughan Bell, Dorothy V. M. Bishop, and Andrew K. Przybylski, "The Debate over Digital Technology and Young People," *BMJ: British Medical Journal* 351 (2015): h3064. Studies so far have tended to consider such factors in isolation rather than as just one component of the highly complex world that digitalization creates (see Chapters 7 and 8).
23. R. M. Nesse, "Natural Selection and the Regulation of Defenses: A Signal Detection Analysis of the Smoke Detector Principle," *Evolution and Human Behavior* 26 (2005): 88–105.
24. Some animals show a decline in fertility with age, but middle-aged human females completely cease ovulating and menstruating. There have been several theories put forward to explain the evolution of menopause, but no

absolute consensus. One argument is that menopause is an epiphenomenon arising simply as a by-product of a highly conserved mammalian pattern. That is, primary oocytes are formed *in utero* only, and thus age over the life course and can accumulate mutations, whereas spermatogenesis continues throughout the male life course. As the human life span increases, the difference between ovarian senility and the life span would become more prominent. L. L. Sievert, "The Evolution of Post-Reproductive Life: Adaptationist Scenarios," in *Reproduction and Adaptation,* ed. C. G. Nicholas Mascie-Taylor and Lyliane Rosetta (Cambridge: Cambridge University Press, 2011).

A second indirect argument is that menopause might reflect a by-product of selection for longevity in males. Shripad D. Tuljapurkar, Cedric O. Puleston, and Michael D. Gurven, "Why Men Matter: Mating Patterns Drive Evolution of Human Lifespan," *PLOS One* 2 (2007): e785. However, most focus has been on arguments for a direct adaptive advantage for menopause. G. C. Williams, in "Pleiotropy, Natural Selection, and the Evolution of Senescence," *Evolution* 11 (1957): 398–411, put forward the "mother hypothesis," according to which menopause is an evolved trade-off because female fertility begins declining well before menopause; modeling shows that the probability of having more surviving children can be increased if the woman stops having children in time to support the development of her youngest child.

A further argument is that a surviving grandmother who no longer actively cares for her own children can promote her own indirect fitness by assisting her daughters in their child rearing. In this view, menopause evolved to promote the grandmother's inclusive fitness. K. Hill and A. M. Hurtado, "The Evolution of Premature Reproductive Senescence and Menopause in Human Females: An Evaluation of the 'Grandmother Hypothesis,'" *Human Nature* 2 (1991): 313–350. There is some observational support for this hypothesis. Cheryl Sorenson Jamison, Laurel L. Cornell, Paul L. Jamison, et al., "Are All Grandmothers Equal? A Review and a Preliminary Test of the 'Grandmother Hypothesis' in Tokugawa Japan," *American Journal of Physical Anthropology* 119 (2002): 67–76. D. P. Shanley and T. B. L. Kirkwood, in "Evolution of the Human Menopause," *Bioessays* 23 (2001): 282–287, used modeling to argue that the strongest fitness effects are supported by a combination of adaptive arguments involving maternal survival, juvenile survival, and grandmaternal assistance. See also D. P. Shanley, R. Sear, R. Mace, et al., "Testing Evolutionary Theories of Menopause," *Proceedings of the Royal Society B* 274 (2007): 2943–2949.

25. Stephen G. Post, Lynn G. Underwood, Jeffrey P. Schloss, and William B. Hurlbut, eds., *Altruism and Altruistic Love: Science, Philosophy, and Religion in Dialogue* (Oxford: Oxford University Press, 2002); Martin A. Nowak and Sarah Coakley, eds., *Evolution, Games, and God: The Principle of Cooperation* (Cambridge, MA: Harvard University Press, 2013).

26. William Hamilton (1936–2000) was one of the most distinguished evolutionary theorists. His arguments on the evolution of altruism were based on concepts of relatedness and thus of what became known as kin selection. Hamilton's rule states that genes will increase in frequency when the genetic relatedness of a recipient to an actor, multiplied by the benefit to the recipient, is greater than the reproductive cost to the actor. He used this to argue for the role of kin selection in the evolution of altruistic behavior. J. B. S. Haldane (1892–1964) was a British polymath, science popularizer, and population geneticist. Another alleged but uncertain quotation of Haldane's is that when asked what he thought could be inferred about the mind of God from the works of his creation, he replied: "An inordinate fondness for beetles." Darwin, too, had spent much of his Cambridge years collecting beetles, and beetle collecting was a major hobby of the emerging naturalists of his era.

27. For a good discussion of this somewhat contentious issue, see David Sloan Wilson, "Reaching a New Plateau for the Acceptance of Multilevel Selection," Evolution Institute (2017), https://evolution-institute.org/focus-article/reaching-a-new-plateau-for-the-acceptance-of-multilevel-selection.

28. Mélanie Salque, Peter I. Bogucki, Joanna Pyzel, et al., "Earliest Evidence for Cheese Making in the Sixth Millennium BC in Northern Europe," *Nature* 493 (2012): 522–525.

29. Dallas M. Swallow, "Genetics of Lactase Persistence and Lactose Intolerance," *Annual Review of Genetics* 37 (2003): 197–219.

30. Nicolás Montalva, Kaustubh Adhikari, Anke Liebert, et al., "Adaptation to Milking Agropastoralism in Chilean Goat Herders and Nutritional Benefit of Lactase Persistence," *Annals of Human Genetics* 83 (2019): 11–22.

31. See Robert Boyd and Peter J. Richerson, *Culture and the Evolutionary Process* (Chicago: University of Chicago Press, 1988); and later discussion in Peter J. Richerson and Robert Boyd, *Not by Genes Alone: How Culture Transformed Human Evolution* (Chicago: University of Chicago Press, 2006); and Robert Boyd, *A Different Kind of Animal: How Culture Transformed Our Species* (Princeton, NJ: Princeton University Press, 2017). Other important papers are L. L. Cavalli-Sforza, M. W. Feldman, K. H. Chen, et al., "Theory and Observation in Cultural Transmission," *Science* 218 (1982): 19–27; and Herbert Gintis, "Gene–Culture Coevolution and the Nature of Human Sociality," *Philosophical Transactions of the Royal Society B: Biological Sciences* 366 (2011): 878–888, which presents gene–culture evolution as a special case of niche construction. We would argue that niche modification is a more distinctive concept, because, at least in modern times, equilibria between biology and culture are not achieved.
32. George Perry, Nathaniel Dominy, Katrina Claw, et al., "Diet and the Evolution of Human Amylase Gene Copy Number Variation," *Nature Genetics* 39 (2007): 1256–1260.
33. For a good summary of the philosophical and practical implications of the false dichotomy between nature (inherited genes) and nurture (environment), see Evelyn Fox Keller, *The Mirage of a Space between Nature and Nurture* (Durham, NC: Duke University Press, 2010).
34. The concept of "social Darwinism" is most often associated with Herbert Spencer (1820–1903), a British polymath whose view of evolution extended well beyond biology to all aspects of human endeavors. He introduced the term *survival of the fittest* in his book *Principles of Biology* (1864) to describe Darwin's theory of natural selection, although his own unifying principle of evolution incorporated use-inheritance and had a strong directionality. At Wallace's suggestion, Darwin introduced the term in the fifth edition of *The Origin of Species* (1869). There is a close relationship between the concepts of directionality as applied to peoples and activities within society, which were manifested in the origin of eugenic arguments and thence in the awful consequences of eugenics in the twentieth century.
35. Franz Boas (1854–1942) was a German-American father of anthropology who was a strong opponent of scientific racism and, being a Darwinist, strongly

argued that biological evolution and cultural development and variation were separate. He objected to the view that there was an evolutionary progression from primitive to advanced cultures. He saw ethnic variation as explained by culture, not by biology, and introduced the concept of cultural relativism. His biological anthropological studies pointed out the importance of environment to growth and skeletal development. See Franz Boas, *The Mind of Primitive Man* (New York: Macmillan, 1911); Boas, *Race, Language, and Culture* (New York: Macmillan, 1940); and Boas, "Changes in the Bodily Form of Descendants of Immigrants," *American Anthropologist* 14 (1912): 530–562. Boas's student Ruth Benedict (1887–1945) and Benedict's student Margaret Mead (1901–1978) continued to develop Boas's arguments for cultural relativism and to argue against scientific racism.

36. Edward O. Wilson, *Sociobiology: The New Synthesis* (Cambridge, MA: Harvard University Press, 1975).
37. Edward O. Wilson, *Consilience: The Unity of Knowledge* (New York: Alfred A. Knopf, 1998).
38. E. O. Wilson's deterministic view is tempered to an extent by his use of the term *epigenetic.* He notes that "epigenetic rules" guide or govern behavior. We might view this as a very modern perception, perhaps softening his gene-centric position, but it appears that his use of *epigenetic* is actually more akin to Waddington's use of the term to reflect switches controlling genetic pathways selected by evolution. Wilson sees the epigenetic processes as setting choices or patterns of behavior that are advantageous for society, such as not mating with siblings, smiling at friends, fearing strangers, and even the grammar of language.
39. The issue of what determines our behavior has other direct impacts. For example, in England the postwar development of "secondary modern" schools for children of lower ability (IQ assessed at the age of eleven) was thought to help equip them for manual trades and domestic skills, preserving the grammar schools for pupils of higher ability. The problem was that IQ is a very poor indicator of many aspects of intellectual ability and is hard to measure. Although such IQ testing was supposed to be independent of home environment, the secondary modern / grammar school dichotomy did much to preserve social class distinctions in the United Kingdom. How-

ever, given this structure, it did promote a degree of social mobility. The persistence of class divisions led to the system being scrapped in the 1970s with the introduction of "comprehensive" schools designed to cater to pupils of all abilities. The discussion thus moved inside the school gates, over whether "streaming" of pupils into classes according to ability was desirable. By this time the concept of a "meritocracy" was deeply entrenched in England—those who commanded higher salaries and positions of power and influence felt entitled to these because of their assumed greater innate abilities. The problem arises, of course, when they in turn become parents. Are they happy to see their children assessed and graded according to a measure of innate ability, which must at least have a component of random genetic variation? Very often they are not. A sort of positive eugenics therefore develops.

40. A. Qiu, T. T. Anh, Y. Li, et al., "Prenatal Maternal Depression Alters Amygdala Functional Connectivity in 6-Month-Old Infants," *Translational Psychiatry* 5 (2015): e508.

41. Examples of Heckman's most insightful work include James J. Heckman, "The Economics, Technology, and Neuroscience of Human Capability Formation," *Proceedings of the National Academy of Sciences* 104 (2007): 13250–13255; J. Heckman and G. Conti, "The Economics of Child Well-Being," in *Handbook of Child Well-Being: Theories, Methods and Policies in Global Perspective,* ed. Asher Ben-Arieh, Ferran Casas, Ivar Frones, and Jill E. Korbin, 363–402 (Dordrecht: Springer, 2013); and James Heckman, Rodrigo Pinto, and Peter Savelyev, "Understanding the Mechanisms through Which an Influential Early Childhood Program Boosted Adult Outcomes," *American Economic Review* 103 (2013): 2052–2086. Heckman also maintains a useful website at https://heckmanequation.org.

42. S. J. Gould and R. C. Lewontin, "The Spandrels of San Marco and the Panglossian Paradigm: A Critique of the Adaptationist Programme," *Proceedings of the Royal Society B* 205 (1979): 581–598.

43. S. J. Gould and E. Vrba, "Exaptation: A Missing Term in the Science of Form," *Paleobiology* 8 (1982): 4–15.

44. A. P. Hendry and M. T. Kinnison, "Perspective: The Pace of Modern Life: Measuring Rates of Contemporary Microevolution," *Evolution* 53 (1999): 1637–1653.

45. G. P. Pfeifer, "Mutagenesis at Methylated CpG Sequences," *Current Topics in Microbiology and Immunology* 301 (2006): 259–281; P. Bateson and P. Gluckman, *Plasticity, Robustness, Development and Evolution* (Cambridge: Cambridge University Press, 2011).

Chapter 5 | Settlements

1. R. I. M. Dunbar, "Coevolution of Neocortical Size, Group Size and Language in Humans," *Behavioral and Brain Sciences* 16 (2010): 681–694. More recently, Dunbar has mentioned two more numbers: an inner core of about five people to whom we devote about 40 percent of our available social time and ten more to whom we devote another 20 percent. All in all, we devote about two-thirds of our time to just fifteen people. Robin Dunbar, "Why Drink Is the Secret to Humanity's Success," *Financial Times,* August 10, 2018, https://www.ft.com/content/c5ce0834-9a64-11e8-9702-5946bae86e6d.
2. One of the best sets of evidence for the butchering of large animals comes from biochemical analysis of protein deposits on the Mahaffy cache, a collection of eighty-three Clovis-era stone tools unearthed on the outskirts of Boulder, Colorado. This showed that the implements were used to butcher ice-age camels and horses that roamed North America until their extinction about 13,000 years ago. The cache was discovered under only about eighteen inches of soil in May 2008 by Brant Turney, who led a landscaping crew working on Patrick Mahaffy's property, and seems to have lain undisturbed for thousands of years. Why the cache was deposited near the Front Range of the Rocky Mountains is not known. It may have been left by seminomadic hunters for seasonal use.
3. Jared Diamond, *Guns, Germs, and Steel* (New York: Norton, 1999).
4. The Danish and French economist Ester Boserup (1910–1999) argued in her book *The Conditions of Agricultural Growth: The Economics of Agrarian Change under Population Pressure* (London: Allen and Unwin, 1965) that population growth drives agricultural output. This countered the contrary view of Malthus that food production determines population growth.
5. Stuart Rojstaczer, Shannon M. Sterling, and Nathan J. Moore, "Human Appropriation of Photosynthesis Products," *Science* 294 (2001): 2549–2552.

6. Douglas J. Kennett, Stephen Plog, Richard J. George, et al., "Archaeogenomic Evidence Reveals Prehistoric Matrilineal Dynasty," *Nature Communications* 8 (2017): 14115.
7. There is a voluminous literature on the evolution of religion, supernatural belief, and ritual. Some useful references are R. A. Hinde, *Why Gods Persist: A Scientific Approach to Religion* (Abingdon, UK: Routledge,1999); Scott Atran, *In Gods We Trust: The Evolutionary Landscape of Religion* (Oxford: Oxford University Press, 2002); Pascal Boyer, *Religion Explained: The Evolutionary Origins of Religious Thought* (New York: Basic Books, 2002); D. S. Wilson, *Darwin's Cathedral: Evolution, Religion and the Nature of Society* (Chicago: University of Chicago Press, 2002); and Ara Norenzayan, *Big Gods: How Religion Transformed Cooperation and Conflict* (Princeton, NJ: Princeton University Press, 2015).
8. See Robert Boyd, *A Different Kind of Animal: How Culture Transformed Our Species* (Princeton, NJ: Princeton University Press, 2017).
9. See P. Gluckman, A. Beedle, T. Buklijas, et al., *Principles of Evolutionary Medicine,* 2nd ed. (Oxford: Oxford University Press, 2016), chap. 5; R. H. Steckel and J. C. Rose, eds., *The Backbone of History: Health and Nutrition in the Western Hemisphere* (New York: Cambridge University Press, 2005); and B. Bogin, *Patterns of Human Growth,* 2nd ed. (Cambridge: Cambridge University Press, 1999).
10. Diamond, *Guns, Germs, and Steel;* Jared Diamond, *Collapse: How Societies Choose to Fail or Succeed* (London: Penguin, 2011).
11. In *The Rise of the West* (Chicago: University of Chicago Press, 1963), the historian William H. McNeill discusses what drives history, arguing that a major force of historical social change for populations has been contact with strangers with new skills and ideas.
12. For more detail on the possible reasons for this migration, see S. G. Ortman, *Winds from the North: Tewa Origins and Historical Anthropology* (Salt Lake City: University of Utah Press, 2012).
13. Wilson, *Darwin's Cathedral.*
14. Harvey Whitehouse and James Laidlaw, eds., *Ritual and Memory: Toward a Comparative Anthropology of Religion* (Walnut Creek, CA: Altamira, 2004).

15. H. Whitehouse, P. François, P. E. Savage, et al., "Complex Societies Precede Moralizing Gods Throughout World History," *Nature* 568 (2019): 226–229.

Chapter 6 | Cities

1. Thomas Malthus (1766–1834) was a cleric and scholar who in 1798 wrote a book (initially published anonymously) entitled *An Essay on the Principle of Population, as it affects the future improvement of society with remarks on the speculations of Mr. Godwin, M. Condorcet, and other writers.* In 1803 he published a second and much lengthier edition, *An Essay on the Principle of Population; or, a view of its past and present effects on human happiness; with an enquiry into our prospects respecting the future removal or mitigation of the evils which it occasions.* The sixth edition, published in 1826, had only minor changes and was the edition Darwin read in 1838, two years after the end of the *Beagle* expedition and while he was formulating his theory. In the first edition Malthus had pointed out that a population would grow faster than food supplies and that this could be dealt with either by active checks such as war or famine, or by preventative checks like delayed marriage, birth control, and celibacy. The second edition responded to a large amount of criticism—for example, from William Hazlitt in *A Reply to the Essay on Population by the Reverend T. R. Malthus* (1807)—and focused more on the latter category of moral constraint, arguing that this would lead to improvement in society. The contemporary debates about the social implications of his argument were intense and linked to the issues regarding the lives of the most disadvantaged sections of the population in Britain. Darwin's understanding of population dynamics shifted after he had read Malthus. Indeed, both Darwin and Wallace considered the Malthusian arguments in developing their concepts of natural selection, but there were differences: Darwin saw competition as largely taking place at the level of individuals, whereas Wallace envisioned it at the level of populations. See Michael Flannery, "Darwin and Wallace Read Malthus Differently, and That Made a Big Difference," *Evolution News & Science Today,* April 20, 2016, https://evolutionnews.org/2016/04/darwin_and_wall/.
2. R. W. Fogel, *The Escape from Hunger and Premature Death, 1700–2100: Europe, America and the Third World* (Cambridge: Cambridge University Press, 2004).

3. To this day the Lord Speaker of the House of Lords in the English Parliament sits on a "woolsack."

4. Daron Acemoglu and James A. Robinson, in *Why Nations Fail: The Origins of Power, Prosperity, and Poverty* (New York: Crown Business, 2012), propose that having either extractive or inclusive economic and political institutions determines whether a country is poor or rich. Inclusive economic institutions provide incentives and opportunities to innovate; extractive systems, where the interests of the elite prevail, do not do so. A striking example of the latter is what is now the Democratic Republic of Congo, with its long history of being ruled by people noted for their immense wealth and greed. More recently, Mobutu Sese Seko, the president from 1965 to 1997, ordered a palace complex to be built at his birthplace. This also contained a specially constructed airport runway to accommodate his fondness for extravagant trips to Europe on chartered planes, including the supersonic Concorde jet.

5. See Gareth Stedman Jones, "In retrospect: *Das Kapital,*" *Nature* 547 (2017): 401–402.

6. The discipline of sociology is often said to have arisen from the writings of Karl Marx along with those of Max Weber and Émile Durkheim. Durkheim insisted that a major *raison d'etre* for the field was the study of the nature of human societies, especially through their institutions, as culturally specific and passed across generations by learning and cultural acquisition. This we can see as a strongly antireductionist position at a time when the field of psychology was developing to propound an essential, biological, and universal basis for "human nature."

   Morgan's magnum opus, *Systems of Consanguinity and Affinity for the Human Family,* was published in July 1871 after many delays (including a major fire at his publisher, the Smithsonian Institution, which could easily have destroyed the manuscript of his life's work, had he submitted it on time). This was the same year as the publication of Darwin's *The Descent of Man.* The influence of one upon the other was perhaps not great. It is more likely that the work in the mid-nineteenth century on the fossil record and early hominins, and acceptance that the history of the world extended much farther back than the supposed biblical time frame, had greater influence. We noted earlier the major contribution of Herbert Spencer, who coined the phrase "survival of the fittest" in his book *Principles of Biology* (1864) after reading Darwin's *The Origin of Species*. Spencer's view of evolution was far more Lamarckian than Darwin's, including the concept that, unlike traits that have fallen into disuse, traits that are used will be inherited. Applying this to social processes led him to suggest that they involved progress from lower to higher forms. He argued that "intuitive" knowledge, which shapes societies in the process, was inherited as part of the shared experience of the species, or at least its component populations by which he inferred "race." So, by the last decades of the nineteenth century, the concepts that would later become incorporated into neo-Darwinism were already in place. The scene was set for the division between biology and culture to be played out, along with the scientific movements of evolutionary psychology, on the one hand, and social or cultural evolution, on the other.

7. E. A. Wrigley, *The Path to Sustained Growth: England's Transition from an Organic Economy to an Industrial Revolution* (Cambridge: Cambridge University Press, 2016).
8. Edward Glaeser, *Triumph of the City: How Our Greatest Invention Makes Us Richer, Smarter, Greener, Healthier, and Happier* (New York: Penguin Press, 2011).
9. See Luís M. A. Bettencourt, José Lobo, Dirk Helbing, et al., "Growth, Innovation, Scaling, and the Pace of Life in Cities," *Proceedings of the National Academy of Sciences* 104 (2007): 7301–7306; and Geoffrey West, *Scale: The Universal Laws of Life, Growth, and Death in Organisms, Cities, and Companies* (New York: Penguin Press, 2017).
10. Robert K. Merton, "The Unanticipated Consequences of Purposive Social Action," *American Sociological Review* 1 (1936): 894–904.
11. In the mid-nineteenth century, people obtained water from town wells and communal pumps, and the germ theory of disease (that diseases are caused and transmitted by microorganisms) had not yet been developed. Instead, the prevailing "miasma" theory pinned the blame for disease on pollution or "bad air." John Snow (1813–1858), often called the first epidemiologist, noted that cases of cholera in the 1854 outbreak in London strongly clustered around one particular street pump, and meticulously gathered evidence from infected and uninfected individuals to link the outbreak to that

pump. It turned out that a woman had dumped water used to wash her infected child's diapers into a leaky cesspool that was located a mere three feet from the street pump. Removing its handle rendered the pump unusable, which was unpopular with the local population but directly prevented any further cases of infection. Yet, despite the immediate effect of this on cholera rates, it took public health officials twelve more years—and the work of others, such as Robert Koch, who developed germ theory—to accept Snow's explanation that disease could be spread through fecal contamination of water.

12. Semmelweis died in an asylum for the insane in 1865, following ridicule and victimization for his theory that the sepsis that affected women in some lying-in wards after childbirth could be prevented by doctors and medical students washing their hands before examining patients—especially if they had just come from dissecting cadavers in the anatomy theater. Semmelweis would have been gratified to see the simple contraption Joseph Lister used in operating rooms beginning in the 1870s, which sprayed a solution of carbolic acid into the air around the surgical site. This was a very effective method of killing bacteria and dramatically reduced mortality following surgical procedures.

13. The German physician Robert Koch (1843–1910) and the French microbiologist Louis Pasteur (1822–1895) are considered to have played the most important roles in the foundations of medical microbiology. Koch is especially well known for establishing four postulates for identifying the causative agent of a disease: that the pathogen should (1) be present in all cases of the disease; (2) be taken from a diseased host and grown in culture; (3) when cultured, cause infection in a healthy animal; and (4) be taken from the new host and confirmed as the same pathogen as the original. This framework proved critical in understanding the microbial basis of disease—albeit with limitations, as it was formulated before the identification of viruses as disease agents that do not fulfill some of the postulates. Pasteur showed that fermentation, and spoilage of beverages, was caused by living microbes rather than by chemical decomposition of organic matter. This spurred him to invent a process to sterilize liquids by mild heat treatment. This process, now known as pasteurization, was first used to extend the shelf life of wine and beer, and is now an indispensable part of nearly all milk production.

14. World Health Organization, *The World Health Report 2001: Mental Health: New Understanding, New Hope* (Geneva: World Health Organization, 2001); Max Roser and Hannah Ritchie, "Mental Health" (2019), accessed January 21, 2019, https://ourworldindata.org/mental-health; R. C. Kessler, Matthias Angermeyer, James C. Anthony, et al., "Lifetime Prevalence and Age-of-Onset Distributions of Mental Disorders in the World Health Organization's World Mental Health Survey Initiative," *World Psychiatry* 6 (2007): 168–176; World Health Organization, *Mental Health Atlas 2017* (Geneva: World Health Organization, 2018).
15. Andreas Heinz, Lorenz Deserno, and Ulrich Reininghaus, "Urbanicity, Social Adversity and Psychosis," *World Psychiatry* 12 (2013): 187–197; Joanne Newbury, Louise Arseneault, Avshalom Caspi, et al., "Why Are Children in Urban Neighborhoods at Increased Risk for Psychotic Symptoms? Findings from a UK Longitudinal Cohort Study," *Schizophrenia Bulletin* 42, no. 6 (2016): 1372–1383.
16. Andrew Jakubowicz, Kevin Dunn, Gail Mason, et al., *Cyber Racism and Community Resilience: Strategies for Combating Online Race Hate* (London: Palgrave Macmillan, 2017).
17. In one of the more controversial observations made to support evolutionary psychological arguments, two distinguished evolutionary psychologists, Martin Daly and Margo Wilson, drew the conclusion from demographic data that it was more likely that a stepfather would kill a young stepchild than that a biological father would kill his biological child; Martin Daly and Margo I. Wilson, "Some Differential Attributes of Lethal Assaults on Small Children by Stepfathers versus Genetic Fathers," *Ethology and Sociobiology* 15 (1994): 207–217. This association has been termed the "Cinderella effect." This observation and other evidence of differential care has been repeated in various countries; see, for instance, Greg A. Tooley, Mari Karakis, Mark Stokes, et al., "Generalising the Cinderella Effect to Unintentional Childhood Fatalities," *Evolution and Human Behavior* 27 (2006): 224–230. Daly and Wilson used an evolutionary argument to support the origin of this behavior, but their argument caused a response that echoed the original sociobiological debates (see Chapter 4), with critics claiming that Daly and Wilson had exaggerated the effect size and discounted confounders such as poverty. M. Daly and M. Wilson, "Discriminative Parental Solicitude:

A Biological Perspective," *Journal of Marriage and Family* 42 (1980): 277–288; Daly and Wilson, "Violence against Stepchildren," *Current Directions in Psychological Science* 5 (1996): 77–80. An extensive literature criticizes their evolutionary interpretation; however, the empirical observations stand, even if the effect size appears to be less than originally claimed and apparent only in very young children. See G. Nobes, G. Panagiotaki, and K. Russell Jonsson, "Child Homicides by Stepfathers: A Replication and Reassessment of the British Evidence," *Journal of Experimental Psychology: General* (2018), https://doi.org/10.1037/xge0000492.

For a useful review of the different approaches to incorporating evolutionary perspectives into studies of human behavior, see K. N. Laland and G. R. Brown, *Sense and Nonsense: Evolutionary Perspectives on Human Behaviour* (Oxford: Oxford University Press, 2002).

Evolutionary psychology is the term used to refer to a strong evolutionary deterministic view of human behavior generally related to the model adopted in Leda Cosmides and John Tooby, "Evolutionary Psychology: New Perspectives on Cognition and Motivation," *Annual Review of Psychology* 64 (2013): 201–229. David J. Buller, in "Evolutionary Psychology: The Emperor's New Paradigm," *Trends in Cognitive Sciences* 9 (2005): 277–283, argued that Cosmides, Tooby, Wilson, and Daly were seeking to explain current behaviors as psychological adaptations rather than giving evolutionary explanations of psychological traits. The latter is closer to the perspective taken by one of the leaders of modern evolutionary medicine and psychiatry, Randolph Nesse. See R. M. Nesse, *Good Reasons for Bad Feelings* (London: Allen Lane, 2019). In turn, it is useful to refer to the four questions formulated by the great ethologist Niko Tinbergen in "On Aims and Methods of Ethology," *Zeitschrift für Tierpsychologie* 20 (1963): 410–433, where he seeks to explain any evolved trait in terms of its biological mechanism, ontogeny, current function, and evolution. Importantly, its current function need not be the same as its function when the trait first evolved.

18. Peter Gluckman and Kristiann Allen, "Understanding Wellbeing in the Context of Rapid Digital and Associated Transformations: Implications for Research, Policy and Measurement: A Discussion Paper," Auckland: International Network for Government Science Advice (2018), https://www.ingsa.org/wp-content/uploads/2018/10/INGSA-Digital-Wellbeing-Sept18.pdf.

Chapter 7 | Online

1. Logarithms enable much easier ways of dealing with complex mathematical problems, such as multiplication and division of large numbers, by simplifying them to involve addition and subtraction. For example 1,000 in logarithmic terms (to base 10) is 10^3, and 1,000,000 is 10^6. Multiplying them together is $10^{(3+6)} = 10^9 = 1{,}000{,}000{,}000$. Large sets of tables were developed so that any number could be expressed as a logarithm.
2. Blaise Pascal (1623–1662) was also a physicist, inventor, and religious philosopher whose name is associated with many well-known ideas. One of these is Pascal's Wager, a philosophical argument that belief in God is the most rational approach: If God does not exist, then belief in God (theism) incurs the relatively small cost of the loss of some pleasures and luxuries; but if God does actually exist, the benefits of theism (eternal life) far outweigh the risks of atheism (eternal damnation).
3. Charles Babbage was a polymath who made important contributions to mathematics, mechanical engineering, and philosophy. His conception of the computing engine in 1834 laid the foundation for modern computer design, and for that he is credited as a true computer pioneer. The mathematician Ada Lovelace recognized that Babbage's mechanical computer could be exploited for purposes far wider than calculations alone, and her derivation of the algorithm for Babbage's machine is widely acknowledged as the first instance of computer programming.
4. Michael Faraday (1791–1867) was a physicist who made major contributions to the study of electricity and magnetism. An often-quoted anecdote, of uncertain veracity, is that when the British politician Benjamin Disraeli asked Faraday what use his discoveries related to electricity might have, Faraday replied, "There is every probability that you will soon be able to tax it." This is considered one of the great defenses of the value of basic research.
5. Coding every number, letter, or symbol as a series of 0s and 1s was the conceptual innovation fundamental to the development of modern computing. This is important because until quantum computers become a realistic proposition, the basis of a computer lies in units that can exist only in one of two states—such as 0 / 1, on / off, or + / –. These binary switches first

comprised vacuum tubes (the size and heat production of which limited the potential power of the early computer), then many more transistors, and now billions of electronic switches on a silicon chip. Such binary codes are not new. Gottfried Leibnitz invented what we consider the modern binary code in 1679. But Chinese and Indian scholars had developed binary systems 2,000 years earlier, and some Polynesians, well before Europeans appeared in the Pacific, also had a binary system. Andrea Bender and Sieghard Beller, "Mangarevan Invention of Binary Steps for Easier Calculation," *Proceedings of the National Academy of Sciences* 111 (2014): 1322–1327.

6. Ray Kurzweil, *The Singularity Is Near: When Humans Transcend Biology* (New York: Penguin, 2005). Others—such as Jamie Bartlett in *The People vs Tech: How the Internet Is Killing Democracy (and How We Save It)* (London: Ebury Press, 2018)—have argued that the real issue is not a technical singularity but a moral singularity, when machines, on our behalf, make decisions that involve moral judgment.

7. The transhumanists are techno-utopians who are seeking technologies that will transcend the limits of the human condition. They seek enhancements of body and mind, and radical life extension, via technologies. Nick Bostrom, "A History of Transhumanist Thought," in *Academic Writing across the Disciplines,* ed. Michael Rectenwald and Lisa Carl (New York: Pearson Longman, 2011); Bostrom, "What Is Transhumanism? (2001), https://nickbostrom.com/old/transhumanism.html; Andrew Pilsch, *Transhumanism: Evolutionary Futurism and the Human Technologies of Utopia* (Minneapolis: University of Minnesota Press, 2017). Indeed, some transhumanists have introduced the term *techno sapiens* with the suggestion that transhumanism could effectively lead to a new species. Francis Fukuyama has argued that the very idea of seeking to have our biological limits overridden by technology is dangerous, because he sees it as being closely linked to extreme libertarianism and detrimental to social equality. Francis Fukuyama, "Transhumanism," *Foreign Policy,* special report, October 23, 2009, https://foreignpolicy.com/2009/10/23/transhumanism/. Others have also objected to the concept on ethical or spiritual grounds. There are more pragmatic issues regarding what is technologically feasible. For a valuable ethical overview, see Wendell Wallach, *A Dangerous Master: How to Keep Technology from Slipping beyond Our Control* (New York: Basic Books, 2015).

8. For an insider's journalistic view of the drivers of some of the key players in platform companies, see Roger McNamee, *Zucked: Waking up to the Facebook Catastrophe* (London: HarperCollins, 2019).
9. Different search engines use different algorithms to list links in an internet search. When Google first started, its approach was different from that of its competitors—its ranking algorithm, called PageRank, was based on assessing both relevance and importance as measured by the number of linkages. This gave Google a marked competitive advantage and market dominance. Its search algorithms are frequently updated and are core to its success. Once a search engine like Google has broader access to information online, its utility to the user is reinforced and there are few incentives to break the effective monopoly.
10. R. Dunbar, *Grooming, Gossip, and the Evolution of Language* (Cambridge, MA: Harvard University Press, 1998).
11. For a journalistic discussion on the role of the platform companies in these recent events, see Bartlett, *The People vs Tech.*
12. Carole Cadwalladr, "The Great British Brexit Robbery: How Our Democracy Was Hijacked," *The Observer,* May 7, 2017, https://www.theguardian.com/technology/2017/may/07/the-great-british-brexit-robbery-hijacked-democracy.
13. Much has been written about the impact of digital transformation on both the concept of privacy and the implications for individuals and societies. Views vary, ranging from arguments that the concept of privacy is important and must be protected but that its nature is evolving, to accepting that privacy as we know it has been irreversibly lost in the transformation. For further discussion of this contested space, see Daniel J. Solove, *The Digital Person: Technology and Privacy in the Information Age* (New York: NYU Press, 2006); and David Houle, *Is Privacy Dead? The Future of Privacy in the Digital Age* (e-book, David Houle and Associates, 2013). Many national human rights commissions have also considered the issues. For instance, New Zealand Human Rights Commission, *Privacy, Data and Technology: Human Rights Challenges in the Digital Age* (Auckland: New Zealand Human Rights Commission, 2018).
14. Adam D. I. Kramer, Jamie E. Guillory, and Jeffrey T. Hancock, "Experimental Evidence of Massive-Scale Emotional Contagion through Social

Networks," *Proceedings of the National Academy of Sciences* 111 (2014): 8788–8790. A week later the journal published a statement of editorial concern over the paper. Inder M. Verma, "Editorial Expression of Concern: Experimental Evidence of Massive-Scale Emotional Contagion through Social Networks," *Proceedings of the National Academy of Sciences* 111 (2014): 10779. There was much agitation on social media about the publication and the failure of an Ethics Committee to take jurisdiction over it.

15. Klaus Schwab, *The Fourth Industrial Revolution* (New York: Crown Business, 2017).

16. Braden Allenby, "Emerging Technologies and the Future of Humanity," *Bulletin of the Atomic Scientists* 71 (2015): 29–38.

17. The increasing digitalization of the workplace has numerous downstream effects, ranging from benefits—such as the ability to work remotely, to engage with clients better, and to automate processes to improve efficiency—to concerns about large-scale replacement of human workers with robots and other AI machinery. See, for example, McKinsey & Company, "Future of Work" (2019), https://www.mckinsey.com/featured-insights/future-of-work; "Future of Work," *Nature* 550 (2017): 315; and Carl Benedikt Frey and Michael A. Osborne, "The Future of Employment: How Susceptible Are Jobs to Computerisation?," *Technological Forecasting and Social Change* 114 (2017): 254–280. This topic is now a core policy issue for many governments, and is well recognized within international bodies such as the European Union and the International Labour Organization.

18. Peter Gluckman and Kristiann Allen, "Understanding Wellbeing in the Context of Rapid Digital and Associated Transformations: Implications for Research, Policy and Measurement: A Discussion Paper," Auckland: International Network for Government Science Advice, 2018, https://www.ingsa.org/wp-content/uploads/2018/10/INGSA-Digital-Wellbeing-Sept18.pdf.

19. Algorithms based on previous crime data have been used to predict where criminal activities may next occur. Facial recognition systems that identify persons of interest in crowds have also been given trial runs, with mixed results. Beyond assessments of the effectiveness of such systems, there is now increasing emphasis that such technologies must be transparent and respect citizen privacy. Policy Connect, "Crime Prevention through Artificial

Intelligence" (2019), https://www.policyconnect.org.uk/appgda/research/crime-prevention-through-artificial-intelligence.

20. Across the globe there are multiple pilot projects under way in which a subset of the population receives a non-means-tested standard amount of money, or the benefits of negative income tax, to assess the feasibility and effectiveness of full-scale implementation. A universal basic income is often proposed as a way to mitigate the potential loss of jobs caused by AI technologies.

21. The United Nations Convention on Certain Conventional Weapons has had meetings regarding the use or control of lethal fully autonomous weapon systems (LAWS). Group of Governmental Experts—Convention on Certain Conventional Weapons, *Report of the 2018 Session of the Group of Governmental Experts on Emerging Technologies in the Area of Lethal Autonomous Weapons Systems* (Geneva: United Nations Office at Geneva, 2018). Some key countries are not supportive of an accord, and so progress has stalled.

22. Lord Rees has commented that compared with Earth's four-billion-year history, it would take machines just a few centuries to "have taken over" from humans, with billions of years on Earth ahead of them. Martin Rees, "Aliens, Very Strange Universes and Brexit—Martin Rees Q&A," The Conversation, April 3, 2017, https://theconversation.com/aliens-very-strange-universes-and-brexit-martin-rees-qanda-75277.

23. Robert Chesney and Danielle Citron, "Deepfakes and the New Disinformation War," *Foreign Affairs,* January / February 2019, https://www.foreignaffairs.com/articles/world/2018-12-11/deepfakes-and-new-disinformation-war.

24. Robert J. Shiller, *The Subprime Solution: How's Today's Global Financial Crisis Happened, and What to Do about It* (Princeton, NJ: Princeton University Press, 2008); Ian Goldin and Chris Kutarna, *Age of Discovery: Navigating the Risks and Rewards of Our New Renaissance* (New York: St Martin's Press, 2016).

25. For example, it is now well established that infants should be put to sleep on their backs rather than on their stomachs or their sides. It was discovered some twenty years ago that this drastically reduces the risk of sudden infant death (cot death), reversing a long-standing view that babies should be placed face down to avoid any risk of inhaling food if they regurgitate.

However, a relatively recent survey of websites advising on infant sleep position found an alarmingly high percentage of sites that advised the wrong position. Matthew Chung, Rosalind P. Oden, Brandi L. Joyner, et al., "Safe Infant Sleep Recommendations on the Internet: Let's Google It," *Journal of Pediatrics* 161 (2012): 1080–1084.e1.

26. The phrase *alternative facts* was first used by Kellyanne Conway, counselor to U.S. president Donald Trump, during an interview on January 22, 2017, in which she defended White House press secretary Sean Spicer's demonstrably false statement about the attendance numbers at Trump's presidential inauguration. Since then the phrase has entered popular parlance to denote the intentional misstatement of reality for political purposes.

27. There have been many books written about "post-truth." For instance, Lee McIntyre, *Post-Truth* (Cambridge, MA: MIT Press, 2018); Matthew d'Ancona, *Post Truth: The New War on Truth and How to Fight Back* (London: Ebury Press, 2017); Evan Davis, *Post Truth: Why We Have Reached Peak Bullshit and What We Can Do about It* (London: Little, Brown, 2017); and Dave Levitan, *Not a Scientist: How Politicians Mistake, Misrepresent and Utterly Mangle Science* (New York: Norton, 2017). The reality is that the phenomenon is not new, but the internet makes it so much easier to broadcast intentionally misleading statements for political or other purposes.

28. The UK Brexit-favoring politician Michael Gove, speaking on BBC in July 2016, said, "I think that the people of this country have had enough of experts from organizations with acronyms saying that they know what is best and getting it consistently wrong, because these people are the same ones who got it consistently wrong." The quote is often abbreviated to the first thirteen words.

29. This is a well-known social psychological phenomenon called biased assimilation, where people with divergent attitudes to a particular issue interpret new information in a manner consistent with their original attitudes. Adam Corner, Lorraine Whitmarsh, and Dimitrios Xenias, "Uncertainty, Scepticism and Attitudes towards Climate Change: Biased Assimilation and Attitude Polarisation," *Climatic Change* 114 (2012): 463–478. The impact is further complicated by the many contextual factors affecting individual worldviews, loyalties, and trust in expert opinion. *Backfire effect* and *belief polarization* are

terms used to describe how this biased assimilation can polarize views further. Jack Zhou, "Boomerangs versus Javelins: How Polarization Constrains Communication on Climate Change," *Environmental Politics* 25 (2016): 788–811; S. L. van der Linden, A. Leiserowitz, S. Rosenthal, and E. Maibach, "Inoculating the Public against Misinformation about Climate Change," *Global Challenges* 1 (2017), https://doi.org/10.1002/gch2.201600008; Stephan Lewandowsky, Klaus Oberauer, and Gilles E. Gignac, "NASA Faked the Moon Landing—Therefore, (Climate) Science Is a Hoax: An Anatomy of the Motivated Rejection of Science," *Psychological Science* 24 (2013): 622–633; Stephan Lewandowsky, Gilles E. Gignac, and Klaus Oberauer, "The Robust Relationship between Conspiracism and Denial of (Climate) Science," *Psychological Science* 26 (2015): 667–670; and John Cook and Stephan Lewandowsky, "Rational Irrationality: Modeling Climate Change Belief Polarization Using Bayesian Networks," *Topics in Cognitive Science* 8 (2016): 160–179. Recent data have shown that the backfire effect may be more limited and that motivated rejection of corrections of myths, particularly those projected in the political context, occurs only for corrections that directly challenge strong attitudes. Michael J. Aird, Ullrich K. H. Ecker, Briony Swire, et al., "Does Truth Matter to Voters? The Effects of Correcting Political Misinformation in an Australian Sample," *Royal Society Open Science* 5 (2018): 180593.

30. While the antagonism in some quarters to vaccination predates the alleged link between measles immunization and autism, the "anti-vax" movement seized on a controversial paper published by Andrew Wakefield and colleagues in *The Lancet* in 1998. Despite the fact that this was only an observational report on twelve autistic children, Wakefield claimed it implied a link between autism and the MMR (measles, mumps, and rubella) vaccine. The paper received widespread publicity, with Wakefield denouncing the safety of the vaccine in press conferences and other interviews. However, the study findings were not reproduced by other groups. Further investigative reporting by a journalist, and then an inquiry launched by the British General Medical Council, found numerous issues with the study and Wakefield himself: including that Wakefield had undisclosed financial conflicts of interest; had performed the study without ethical approval; had unnecessarily subjected children to invasive medical procedures; and had egregiously falsified data. The paper was retracted two years later, and Wakefield

was found guilty of fraudulent behaviors associated with the study and was struck off the medical register. However, he continued to promote the anti-vaccination agenda upon moving to the United States, and many uninformed celebrities have climbed onto the anti-vax bandwagon. The whole episode is thought to be the major reason for the significant decline in MMR vaccinations, which led to a resurgence of measles. This is despite extensive work done since that strongly refutes any causal relationship between MMR vaccination and autism. See, for example, Institute of Medicine Immunization Safety Review Committee, *Immunization Safety Review: Vaccines and Autism* (Washington, DC: National Academies, 2004).

31. Churchill is often misquoted, but in a speech in the House of Commons in 1947 he said, "No one pretends that democracy is perfect or all-wise. Indeed, it has been said that democracy is the worst form of government except all those other forms that have been tried from time to time."

32. Statistics New Zealand, "Integrated Data Infrastructure," https://www.stats.govt.nz/integrated-data/integrated-data-infrastructure/.

33. P. Gluckman, "Using Evidence to Inform Social Policy: The Role of Citizen-Based Analytics: A Discussion Paper," Auckland: Office of the Prime Minister's Chief Science Advisor, 2017, https://www.pmcsa.org.nz/wp-content/uploads/17-06-19-Citizen-based-analytics.pdf.

34. Data Futures Partnership, "Our Data, Our Way: What New Zealand People Expect from Guidelines for Data Use and Sharing: Findings from Public Engagement," February / March 2017, Wellington: Data Futures Partnership, https://trusteddata.co.nz/massey_our_data_our_way.pdf.

35. This regulation is a binding, legislative act that applies to organizations that process data from individuals within the European Union and the European Economic Area. It requires that appropriate measures for data protection be implemented during the collection and handling of personal data. These measures include data anonymization, obtaining informed consent from data subjects, and upholding their rights in accessing or erasing their data. See European Commission, "2018 Reform of EU Data Protection Rules," https://ec.europa.eu/commission/priorities/justice-and-fundamental-rights/data-protection/2018-reform-eu-data-protection-rules_en.

36. Clare Wilson, "Serial Killer Suspect Identified Using DNA Family Tree Website," *New Scientist,* April 27, 2018, https://www.newscientist.com/article/2167554-serial-killer-suspect-identified-using-dna-family-tree-website/.

37. Shoshana Zuboff, *The Age of Surveillance Capitalism: The Fight for a Human Future at the New Frontier of Power* (New York: Public Affairs, 2019).

38. Organisation for Economic Co-operation and Development, *Children and Young People's Mental Health in the Digital Age: Shaping the Future* (Paris: OECD, 2018); Martin P. Paulus, Lindsay M. Squeglia, Kara Bagot, et al., "Screen Media Activity and Brain Structure in Youth: Evidence for Diverse Structural Correlation Networks from the ABCD Study," *NeuroImage* 185 (2019): 140–153; Sheri Madigan, Dillon Browne, Nicole Racine, et al., "Association between Screen Time and Children's Performance on a Developmental Screening Test," *JAMA Pediatrics* (2019), https://doi.org/10.1001/jamapediatrics.2018.5056.

Chapter 8 | Cost

1. P. D. Gluckman and M. A. Hanson, *Mismatch: Why Our World No Longer Fits Our Bodies* (Oxford: Oxford University Press, 2006).

2. P. D. Gluckman and M. A. Hanson, *Fat, Fate and Disease: Why Exercise and Diet Are No Longer Enough* (Oxford: Oxford University Press, 2012).

3. R. C. W. Ma, J. C. N. Chan, W. H. Tam, et al., "Gestational Diabetes, Maternal Obesity and the NCD Burden," *Clinical Obstetrics and Gynecology* 56 (2013): 633–641.

4. P. D. Gluckman, M. A. Hanson, and H. G. Spencer, "Predictive Adaptive Responses and Human Evolution," *Trends in Ecology & Evolution* 20 (2005): 527–533; Patrick Bateson, Peter Gluckman, and Mark Hanson, "The Biology of Developmental Plasticity and the Predictive Adaptive Response Hypothesis," *Journal of Physiology* 592 (2014): 2357–2368.

5. Steven Sloman and Philip Fernbach, *The Knowledge Illusion: Why We Never Think Alone* (New York: Riverhead Books, 2017). This important book points out that we store knowledge as a collective group rather than as individuals, but that as a result, peer pressure can have major influences on what the group thinks, and therefore on what we ourselves think.

6. Calestous Juma, *Innovation and Its Enemies: Why People Resist New Technologies* (Oxford: Oxford University Press, 2016).

7. The precautionary principle is increasingly codified in agreements and legislation, but its arbitrary nature further confounds adaptive management. Gary E. Marchant, "From General Policy to Legal Rule: Aspirations and Limitations of the Precautionary Principle," *Environmental Health Perspectives* 111 (2003): 1799–1803. It is made more difficult by the multiple interpretations of what it means when reduced to practice. Absolute safety or zero risk is impossible to prove, yet that is often what advocates wish the principle to state. In its original meaning it was intended to refer to adaptive regulation, which would be adjusted in light of new, emerging information that changed the assessment of risk. However, most uses of the precautionary principle in regulation have not taken an adaptive approach.

8. P. Gluckman, "New Technologies and Social Consensus," speech, Auckland: Office of the Prime Minister's Chief Science Advisor, 2016, https://www.pmcsa.org.nz/wp-content/uploads/Discussion-of-Social-Licence.pdf.

9. For useful summaries and explanations of the different types of gene editing and gene modification, see Scientific Advice Mechanism High Level Group of Scientific Advisors, *New Techniques in Agricultural Biotechnology: Explanatory Note 02/2017* (Brussels: European Commission, 2017); and Royal Society (London), *GM Plants: Questions and Answers* (London: Royal Society, 2016). Genetic modification is generally taken to mean the insertion of a whole gene from another species (transgenesis) or another copy of a gene from the same species (cisgenesis) into the genome. The technology caused much debate and opposition when it was introduced in the 1980s, and debate and controversy continue to this day, with claim and counterclaim, despite considerable scientific consensus on its safety, with major reviews conducted by respected scientific academies. For example: National Academies of Sciences, Engineering, and Medicine, *Genetically Engineered Crops: Experiences and Prospects* (Washington, DC: National Academies Press, 2016); and David Baulcombe, Jim Dunwell, Jonathan Jones, et al., *GM Science Update: A Report to the Council for Science and Technology,* 2014, https://assets.publishing.service.gov.uk/government/uploads/system/uploads/attachment_data/file/292174/cst-14-634a-gm-science-update.pdf.

Gene editing covers a variety of more precise ways of modifying the genome, essentially by manipulating one or a small number of nucleotides to change the regulation or expression of a gene. In that sense, it has been argued, it is essentially a form of nonrandom rather than random mutagenesis. There are widely disparate views across countries as to whether this should be considered under the same legal classification as genetic modification. The argument put forward by some regulators is that because the same change might happen naturally, it should not be regulated in the same way as genetic modification has been. For a review, see Academy of Science of South Africa, "The Regulatory Implications of New Breeding Techniques," Pretoria, 2017, http://doi.org/10.17159/assaf.2016/0011. For an interesting perspective on the political and related debates on GMOs, see Mark Lynas, *Seeds of Science: Why We Got It So Wrong on GMOs* (London: Bloomsbury Sigma, 2018). There is a strong scientific consensus that gene editing of human germlines is premature. For example, Nuffield Council on Bioethics, *Genome Editing and Human Reproduction: Social and Ethical Issues* (London: Nuffield Council on Bioethics, 2018); National Academies of Sciences, Engineering, and Medicine, *Second International Summit on Human Genome Editing: Continuing the Global Discussion: Proceedings of a Workshop—in Brief* (Washington, DC: National Academies Press, 2019).

10. P. Gluckman, *Making Decisions in the Face of Uncertainty: Understanding Risk: Part 1* (Auckland: Office of the Prime Minister's Chief Science Advisor, 2016), https://www.pmcsa.org.nz/wp-content/uploads/PMCSA-Risk-Series-Paper-1_final_2.pdf; P. Gluckman, *Making Decisions in the Face of Uncertainty: Understanding Risk: Part 2: Risk Perception, Communication and Management* (Auckland: Office of the Prime Minister's Chief Science Advisor, 2016), https://www.pmcsa.org.nz/wp-content/uploads/PMCSA-Risk-paper-2-Nov-2016-.pdf.

11. Jonathan St. B. T. Evans, "In Two Minds: Dual-Process Accounts of Reasoning," *Trends in Cognitive Sciences* 7 (2003): 454–459; Daniel Kahneman, *Thinking, Fast and Slow* (New York: Farrar, Straus and Giroux, 2011).

12. Gerd Gigerenzer, *Reckoning with Risk: Learning to Live with Uncertainty* (London: Penguin, 2003).

13. See https://www.ingsa.org/.

14. Peter Gluckman and Kristiann Allen, "Understanding Wellbeing in the Context of Rapid Digital and Associated Transformations: Implications for Research, Policy and Measurement: A Discussion Paper," Auckland: International Network for Government Science Advice, 2018, https://www.ingsa.org/wp-content/uploads/2018/10/INGSA-Digital-Wellbeing-Sept18.pdf.
15. See Pia R. Britto, Stephen J. Lye, Kerrie Proulx, et al., "Nurturing Care: Promoting Early Childhood Development," *Lancet* 389 (2017): 91–102. The resulting WHO initiative has been very influential. World Health Organization, *Nurturing Care for Early Childhood Development: A Framework for Helping Children Survive and Thrive to Transform Health and Human Potential* (Geneva: World Health Organization, 2018).
16. Kathleen M. Griffiths and Helen Christensen, "Internet-Based Mental Health Programs: A Powerful Tool in the Rural Medical Kit," *Australian Journal of Rural Health* 15 (2007): 81–87.
17. Karl Popper, *The Logic of Scientific Discovery* (Abingdon, UK: Routledge, 2002).
18. United Nations Department of Economic and Social Affairs, Population Division, *World Population Prospects: The 2017 Revision, Key Findings and Advance Tables,* Working Paper No. ESA/P/WP/248 (New York: United Nations, 2017).
19. In evolutionary terms, fitness is defined in terms of reproductive success. In practice, this means that it is the interaction between genes and the environment that determine fitness in a particular context, although the extended evolutionary arguments would consider epigenetic factors as modifiers of the genetic component in the interaction. For humans, though, cultural factors play an increasingly key role in determining reproductive success—birth control and the expanding use of assisted reproductive technologies being the primary factors for many in determining reproductive outcomes. This creates some practical difficulties in interpreting studies of human fitness and biological drivers. For example, to study evidence of recent changes in the genome, should one focus on natural fertility populations that are not practicing birth control or using reproductive technologies? This was the approach taken in Ken R. Smith, Heidi A. Hanson, Geraldine P. Mineau, et al., "Effects of *BRCA1* and *BRCA2* Mutations on Female Fertility," *Proceedings*

of the Royal Society B 279 (2011): 1389–1395. Or should one, as many evolutionary biologists would argue, accept that culture itself is part of our evolutionary path and thus include reproductive technologies in thinking about contemporary human evolution? From this position, Byars et al. suggest from studies of a multigeneration cohort that there is ongoing significant selection on height, weight, and age at first birth and menopause. Sean G. Byars, Douglas Ewbank, Diddahally R. Govindaraju, et al., "Natural Selection in a Contemporary Human Population," *Proceedings of the National Academy of Sciences* 107 (2010): 1787–1792. The complexities are discussed in Stephen C. Stearns, Sean G. Byars, Diddahally R. Govindaraju, et al., "Measuring Selection in Contemporary Human Populations," *Nature Reviews Genetics* 11 (2010): 611–622. As population geneticists seek evidence of genomic evolution in humans, the issues of how to interpret the data and deal with rapidly changing environmental and cultural factors are a challenge for studies of contemporary evolution. Felix C. Tropf, Gert Stulp, Nicola Barban, et al., "Human Fertility, Molecular Genetics, and Natural Selection in Modern Societies," *PLOS One* (2015), https://doi.org/10.1371/journal.pone.0126821.

20. In an important essay, science policy scholar Dan Sarewitz takes a far more utilitarian view of the scientific enterprise than does Vannevar Bush, and points out the key role of new technologies in driving science forward. Daniel Sarewitz, "Saving Science," *New Atlantis,* Spring / Summer 2016, 5–40.

21. Garrett Hardin, "The Tragedy of the Commons," *Science* 162 (1968): 1243–1248.

22. *Meiotic gene drive* refers to one copy of a gene being passed on to offspring more than the expected 50 percent of the time. Although this is relatively rare in animals, it is at least conceptually possible with genetic editing to use the mechanism that drives this phenomenon to regulate the passage of a desirable mutant gene more rapidly through a population than would otherwise occur. It most likely will be used first in insect control. For a review, see V. M. Macias, J. R. Ohm, and J. L. Rasgon, "Gene Drive for Mosquito Control: Where Did It Come from and Where Are We Headed?," *International Journal of Environmental Research and Public Health* 14 (2017): 1006–1035.

23. For example, soil samples from twenty US native prairies that have not been subject to human or animal grazing activity in historical times contained

bacteria resistant to two common broad-spectrum antibiotics. Nearly half of the bacteria isolated from the soil showed resistance to two or more antibiotics. Lisa M. Durso, David A. Wedin, John E. Gilley, et al., "Assessment of Selected Antibiotic Resistances in Ungrazed Native Nebraska Prairie Soils," *Journal of Environmental Quality* 45 (2016): 454–462. See also William Calero-Cáceres and Maite Muniesa, "Persistence of Naturally Occurring Antibiotic Resistance Genes in the Bacteria and Bacteriophage Fractions of Wastewater," *Water Research* 95 (2016): 11–18.

24. *Cognitive bias,* a concept introduced by Kahneman and Tversky, is an important concept in the psychology of decision making. Daniel Kahneman and Amos Tversky, "Subjective Probability: A Judgment of Representativeness," *Cognitive Psychology* 3 (1972): 430–454. The field describes the inherent biases that come into play when individuals make decisions, including "confirmation bias," where individuals take more notice of information that coincides with their prior beliefs. For a popular science treatment of cognitive biases, see Kahneman, *Thinking, Fast and Slow.*

25. Thomas Dietz, "Bringing Values and Deliberation to Science Communication," *Proceedings of the National Academy of Sciences* 110, suppl. 3 (2013): 14081–14087.

26. Catherine Clifford, "Elon Musk: Mark My Words—A.I. Is Far More Dangerous than Nukes," CNBC—Make It, March 13, 2018, https://www.cnbc.com/2018/03/13/elon-musk-at-sxsw-a-i-is-more-dangerous-than-nuclear-weapons.html.

27. The number of reports analyzing the future of work is increasing. Examples: McKinsey & Company, *Future of Work,* 2019, https://www.mckinsey.com/featured-insights/future-of-work; "Future of Work," *Nature* 550 (2017): 315; European Group on Ethics in Science and New Technologies, *Proceedings of the Open Round Table on the Future of Work, February 5, 2018* (Brussels: European Commission, 2018); International Labour Organization, *Inception Report for the Global Commission on the Future of Work* (Geneva: ILO, 2017).

Chapter 9 | Future

1. Mountain Gorilla Conservation Fund, "Questions about Gorillas," accessed January 17, 2019, http://www.saveagorilla.org/60-Questions.html.

2. Gorilla Doctors, "About Us," accessed January 17, 2019, http://www.gorilladoctors.org/about-us.
3. S. Jay Olshansky, Douglas J. Passaro, Ronald C. Hershow, et al., "A Potential Decline in Life Expectancy in the United States in the 21st Century," *New England Journal of Medicine* 352 (2005): 1138–1145.
4. For an ethicist's perspective, see Wendell Wallach, *A Dangerous Master: How to Keep Technology from Slipping beyond Our Control* (New York: Basic Books, 2015); and Nigel Shadbolt and Roger Hampson, *The Digital Ape: How to Live (in Peace) with Smart Machines* (Melbourne: Scribe, 2018).
5. In their book *The Major Transitions in Evolution* (Oxford: Oxford University Press, 1995), John Maynard Smith and Eörs Szathmáry pointed out that a number of major transitions occurred during evolution, each of which expanded the evolutionary possibilities and potential for biological complexity. These transitions included the encapsulation of replicating molecules, the development of chromosomes, the transition from RNA to DNA as the coding molecule, the formation of eukaryotes, the evolution of sexual reproduction, the development of multicellularity, the development of social animals, the development of language, and cultural evolution.
6. M. M. Ali and D. S. Dwyer, "Estimating Peer Effects in Sexual Behavior among Adolescents," *Journal of Adolescence* 34 (2011): 183–190.
7. D. D. Luxton, J. D. June, and J. M. Fairall, "Social Media and Suicide: A Public Health Perspective," *American Journal of Public Health* 102-S2 (2012): S195–S200.
8. The concept of the Westphalian state, where national powers respect the jurisdictional integrity of other countries and do not interfere in their domestic policies, is derived from the implications of the Peace of Westphalia (1648), which ended the Thirty Years War, a part of the lengthy conflict over religious beliefs in Europe that cost millions of lives. Peace necessitated a series of treaties involving many countries in Europe. This remains the basis of much international law today, but it does not directly address issues of human rights or democracy, although promotion of these is increasingly implied when the term is used.
9. Paul Cairney, *The Politics of Evidence-Based Policy Making* (London: Palgrave MacMillan, 2016); P. Gluckman, "The Role of Evidence and Expertise in

Policy-Making: The Politics and Practice of Science Advice," *Journal and Proceedings of the Royal Society of New South Wales* 1515 (2018): 91–101; P. Gluckman, "Can Science and Science Advice Be Effective Bastions against the Post-Truth Dynamic?," speech, 2017, Auckland, Office of the Prime Minister's Chief Science Advisor, https://www.pmcsa.org.nz/wp-content/uploads/17-10-18-UCL-speech.pdf.

10. Ethicists including Gary Marchant and Wendell Wallach propose Governance Coordinating Committees to allow for adaptive management of new technologies, their point being that "soft law" is better for managing emerging and evolving technologies. In contrast, hard regulation can be inhibitory, difficult to change, and unresponsive to new knowledge. Gary E. Marchant and Wendell Wallach, "Governing the Governance of Emerging Technologies," in *Innovative Governance Models for Emerging Technologies,* ed. Gary E. Marchant, Kenneth W. Abbott, and Braden Allenby (Gloucestershire: Edward Elgar, 2013), 136–152.

11. Paulo Freire (1921–1997) was a Brazilian educator and philosopher. His book *Pedagogy of the Oppressed* (New York: Continuum, 2000) was highly influential in developing the school of critical pedagogy.

12. Peter Gluckman and Kristiann Allen, "Understanding Wellbeing in the Context of Rapid Digital and Associated Transformations: Implications for Research, Policy and Measurement: A Discussion Paper," Auckland, International Network for Government Science Advice, 2018, https://www.ingsa.org/wp-content/uploads/2018/10/INGSA-Digital-Wellbeing-Sept18.pdf.

13. The so-called scientific method that is the basis of Western science combines both theory and empiricism. It is often assumed to have been developed by Francis Bacon, but elements of it were already present in non-European culture. The Arab physicist Ibn al-Haytham (Alhazen) used experimentation to obtain results, combining inductive argument, observations, and experiments in his *Book of Optics* (1021). He emphasized the role of empiricism and considered induction to be the basic requirement for science. Francis Bacon (1561–1626) was a British philosopher and statesman who argued in his book *Novum Organum* (1620) for the need to combine inductive reasoning and observation to advance knowledge. He himself was not an experimenter, but when The Royal Society was founded in 1660, scholars such as Robert

Hooke (1635–1703) were early exponents of the Baconian cycle—that is, the process of moving from observation to hypothesis to experiment to interpretation and hypothesis refinement.

14. Vannevar Bush, *Science: The Endless Frontier: A Report to the President* (Washington, DC: U.S. GPO, 1945), https://www.nsf.gov/od/lpa/nsf50/vbush1945.htm.
15. A first-order interaction considers how two factors, A and B, interact with each other. A second-order interaction considers how a third factor, C, might interact with both A and B and thus alter the interaction between them.
16. *Post-normal science* is a term introduced by Funtowicz and Ravetz to describe science where "facts are uncertain, values in dispute, stakes high and decisions urgent." Silvio O. Funtowicz and Jerome R. Ravetz, "Science for the Post-Normal Age," *Futures* 25 (1993): 739–755. Much of the science interacting with society and policymakers is of this nature. Because of the interface between science and disputed values, there is a danger that science can be used as a proxy for debates that are value-based. For further discussion, see Roger A. Pielke Jr., *The Honest Broker* (Cambridge: Cambridge University Press, 2007). But equally, science is compromised by proceeding without considering the societal perspectives. Hence, post-normal science endorses the importance of the concepts of extended peer review, co-design, and co-production.
17. Pokémon Go is an augmented reality game that became extremely popular for a few months after its release in 2016.
18. United Nations, "Sustainable Development Goals: Knowledge Platform," accessed January 17, 2019, https://sustainabledevelopment.un.org/?menu=1300.
19. For further discussion of the potential of technology-driven social development, see Shadbolt and Hampson, *The Digital Ape;* and Max Tegmark, *Life 3.0: Being Human in the Age of Artificial Intelligence* (New York: Knopf, 2017).
20. Stephen Jay Gould, in his book *Wonderful Life: The Burgess Shale and the Nature of History* (New York: W. W. Norton, 1990), uses the metaphor of the "tape of life," arguing that evolution relies on random mutations and events and therefore that replaying the tape would lead, stochastically, to different outcomes. However, more-recent work suggests that there is some level of

bias in evolutionary processes. See, for example, Alexander E. Lobkovsky and Eugene V. Koonin, "Replaying the Tape of Life: Quantification of the Predictability of Evolution," *Frontiers in Genetics* 3 (2012): 246. Others have pointed out the inevitable convergence, at least in phenotype, that is found in common environments; for example, whales and fish have phenotypic similarities despite being evolutionarily very distinct.

21. The prehistory of our species is complex and is still being resolved with new genetic and paleontological discoveries. Modern *Homo sapiens* evolved at least 150,000 years ago, but recent discoveries suggest that our species may be at least 300,000 years old. Julia Galway-Witham and Chris Stringer, "How Did *Homo sapiens* Evolve?," *Science* 360 (2018): 1296–1298. At the early stages of our divergence from other *Homo* species, there was interbreeding in various populations with both Neanderthals and Denisovans. Qiaomei Fu, Mateja Hajdinjak, Oana Teodora Moldovan, et al., "An Early Modern Human from Romania with a Recent Neanderthal Ancestor," *Nature* 524 (2015): 216–219; Viviane Slon, Fabrizio Mafessoni, Benjamin Vernot, et al., "The Genome of the Offspring of a Neanderthal Mother and a Denisovan Father," *Nature* 561 (2018): 113–116. It has been suggested that such interbreeding may have played a role in the adaptive capacities of humans in different environments. Kwang Hyun Ko, "Hominin Interbreeding and the Evolution of Human Variation," *Journal of Biological Research* (Thessaloniki, Greece) 23 (2016), 17.

22. R. Scott Spurlock, *Man on His Nature,* Gifford Lectures, https://www.giffordlectures.org/lectures/man-his-nature.

23. Charles Scott Sherrington, "Conflict with Nature," in *Man on His Nature,* ed. Charles Scott Sherrington (Cambridge: Cambridge University Press, 2009; orig. ed., 1940), 359–404.

24. Charles Scott Sherrington, "The Wisdom of the Body," in Sherrington, *Man on His Nature,* 103–136.

ACKNOWLEDGMENTS

THIS BOOK HAS BEEN written over the past two years and reflects our interactions with many scholars, practitioners, and policymakers over many more years. And much results from our partnerships with many collaborators in both our experimental and theoretical work. Both of us have spent much of our professional lives researching human development, using a range of approaches from experimental to epidemiological and from behavioral to molecular. We are enormously indebted to the many collaborators, fellows, and students who have worked with us, and to the many others who have critiqued our work at conferences or through reviewing our papers or grant applications—even if we did not always agree with their comments.

About twenty years ago, stimulated by the ideas of David Barker, Nick Hales, and Pat Bateson, we started to engage seriously with the discipline of evolutionary biology and, linked to it, that of evolutionary developmental biology. We began to consider how these fields might give insights into current challenges in human health. In time that led us to produce a textbook on evolutionary medicine. A new set of colleagues emerged. Many will not realize how much our interactions with them have contributed to this book. We hope that by reducing complex, and sometimes controversial, ideas in evolutionary biology to a simple and accessible narrative we have not misrepresented their ideas.

More recently we have both engaged with public policy—Mark primarily with national and global health policy, and Peter more generally with the science-policy nexus both as science advisor to three prime ministers of New Zealand and as chair of INGSA. In these roles there have been opportunities to have an extraordinary range of discussions with politicians, policymakers, technologists, entrepreneurs, and public bodies about the issues of technological innovation and development, in the broadest sense of the word.

Thus, in acknowledging the people named below, it is inevitable that we will have accidentally omitted some who have contributed to the ideas in this book. With apology for any such omissions, we would like to acknowledge those who have helped us, consciously or unconsciously:

In developmental sciences: Keith Godfrey, the late David Barker, Cyrus Cooper, Karen Lillycrop, Rohan Lewis, Caroline Fall, Hazel Inskip, Lucy Green, Kirsten Poore, Felino Cagampang, Karen Temple, Philip Calder, Christopher Torrens, Nicholas Harvey, Mary Barker, Judith Eckert, Tom Fleming, Chandni Jacob, Katherine Woods Townsend, Richard Oreffo, Christopher Byrne and Nicola Englyst (Southampton, UK); Terrence Forrester (Kingston); Michael Meaney, Chong Yap Seng, Ciaran Forde, Neerja Kernani and the GUSTO team (Singapore); Wayne Cutfield, Allan Sheppard (Auckland); Torvid Kiserud, Guttorm Haugen (Bergen and Oslo); Carlos Blanco (Maastricht); Gerhard Visser (Utrecht); Anibal Llanos, Maria Seron-Ferre, Paula Casanello, Francisco Mardones (Santiago); Ronald Ma (Hong Kong); Shane Norris (Johannesberg); Jens Aagaard-Hansen (Copenhagen); Flavia Bustreo and Nigel Rollins (Geneva); the late Nick Hales, Dino Giussani and Sue Ozanne (Cambridge); Ray Noble, Lucilla Poston, Neena Modi, Paul Taylor, Donald Peebles, Eric Jauniaux, Anna David and Charles Rodeck (London); Denis Noble (Oxford); Deborah Sloboda (Hamilton, Canada), John Challis

(Vancouver), Steve Matthews (Toronto); Michael Ross (Los Angeles); John Newnham (Perth).

In evolutionary biology and evolutionary medicine: the late Sir Patrick Bateson (Cambridge); Randy Nesse (Arizona); Peter Ellison (Harvard); Gillian Bentley (Durham, UK); Harvey Whitehouse (Oxford); Hamish Spencer (Otago); Rob Foley (Cambridge); Carl Bergstrom (Seattle); Chris Kuzawa (Chicago); Paul Griffiths and Steve Simpson (Sydney); Tony Pleasants (Hamilton, New Zealand); Graeme Wake, Alan Beedle, and Tatjana Buklijas (Auckland).

In public policy, the social sciences, humanities, and related areas: Kristiann Allen, Anne Bardsley, Stuart McNaughton, Ian Lambie (Auckland); Stephen Goldson (Christchurch); Richie Poulton (Dunedin); Julie Maxton (London); Dirk Pilat and Andy Wyckoff (OECD, Paris); Paul Cairney (Stirling), Ann Mettler (Brussels), Heide Hackmann (Paris); Chris Tyler (London); Sir Partha Dasgupta (Cambridge); James Wilsdon (Sheffield); Sheila Jasanoff (Harvard); Vladimír Šucha and David Mair (Brussels); Vaughan Turekian (Washington); Matthias Kaiser and the Bergen team of post-normal science scholars; Jerry Ravetz (Oxford); Ruth Müller and Michael Penkler (Munich); and Regien Biesma (Groningen). In the digital space, we are grateful to Sir Alan Wilson, Gila Sacks (London), Yuko Harayama (Tokyo), Andrew Chen (Auckland), Vince Galvin, Liz Macpherson (Wellington) and to all those who took part in discussions through the Science Policy Exchange in Auckland and Wellington and in the INGSA workshop in London.

Mark is grateful to Jack Hanson for insights into archaeology and urban systems. Peter is grateful to Judy, Michael Heymann, and Deborah Port for being fellow gorilla watchers.

We owe a particular debt to Dr. Felicia Low in Auckland. She has been a co-author and collaborator on many of our papers in developmental and evolutionary sciences over many years, and she helped us enormously with research for this book, particularly that encompassed in the endnotes. Carlos Blanco and Michael Heymann made helpful comments on the penultimate draft of the manuscript.

As ever, we are grateful to Donald Winchester of Watson Little and Co., London, for his good sense and advice. Janice Audet at Harvard University Press accompanied us on the book's journey from concept to publication, always ready to give help and support. Wendy Nelson and colleagues at the Press made painstaking edits to the manuscript. David Sloan Wilson and two anonymous reviewers provided invaluable feedback on the manuscript.

Jane Kitcher (Southampton, UK) and Megan Stünzner (Auckland) patiently typed several drafts of this material and, as ever, provided help and support in numerous ways.

We must thank our institutions, the Universities of Auckland and Southampton, for supporting our collaboration over many years. The British Heart Foundation has also generously supported Mark's work for many years.

INDEX